Palmer C. (Palmer Chamberlain) Ricketts

History of the Rensselaer Polytechnic Institute

1824-1894

Palmer C. (Palmer Chamberlain) Ricketts

History of the Rensselaer Polytechnic Institute
1824-1894

ISBN/EAN: 9783743686991

Printed in Europe, USA, Canada, Australia, Japan

Cover: Foto ©berggeist007 / pixelio.de

More available books at **www.hansebooks.com**

Observatory.   Ranken House.   Laboratory.   Main Building.   Gymnasium.

RENSSELAER POLYTECHNIC INSTITUTE.   1894.

HISTORY

OF THE

RENSSELAER POLYTECHNIC INSTITUTE

1824—1894

BY

PALMER C. RICKETTS

New York
JOHN WILEY AND SONS
53 EAST TENTH STREET
1895

TO THE MEMORY

OF

Stephen Van Rensselaer

AND

Amos Eaton.

PREFACE.

HAVING recently been compelled to write several brief historical sketches of the Institute the writer became interested in its early history. In preparing these narratives he found the official publications giving the characteristics of the School at the time of its foundation to have become very rare. In fact, very few of them antedating 1840 are known to be in existence. For these reasons he determined to expand the sketches and publish a short history of the Institution which should consist largely of a description of the development of its curriculum.

The student of the history of education will recognize the importance of an account of the early methods of instruction pursued in an institution which was, at once, the first School of Science and the first School of Civil Engineering to be established in any English-speaking country, and if the conceded originality of these methods be also considered it is believed that no excuse for the appearance of this somewhat condensed narrative will be thought necessary.

Interesting information has been obtained from

the recently discovered original minutes of the board of trustees for the twenty-five years immediately following the founding of the School, which were believed to have been destroyed in the fire of 1862, and the thanks of the writer are due the President and Secretary of the present board for placing at his disposal the minutes covering the period from 1862 until the present time.

The author is also under obligation to Professor Henry B. Nason for the loan of a number of the early circulars, to A. J. Weise, Esq., for the picture of the Van Der Heyden mansion ; to James Irving, Esq., for that of the building on the Infant School Lot, and to Professor William G. Raymond for the two photographs from which the pictures showing railroad and hydrographic work of students were taken. The Bibliography at the end of the last chapter shows other sources whence information has been obtained.

P. C. R.

RENSSELAER POLYTECHNIC INSTITUTE,
    TROY, N. Y., January 1, 1895.

CONTENTS.

vii

ILLUSTRATIONS.

x

HISTORY

OF THE

RENSSELAER POLYTECHNIC INSTITUTE.

CHAPTER I.

THE FOUNDATION OF THE SCHOOL.

(At the beginning of this century the study of the physical sciences in the United States was in its infancy. All branches were included under the terms Natural Philosophy and Natural History. Their meaning was not well defined, although under the latter was generally included all of what was then known of astronomy, physics, chemistry and geology. Scarcely any provision was made for scientific instruction in any of the colleges of the country. Astronomy, physics, chemistry and botany had indeed been taught during the preceding century in a few institutions of learning, a department of Mathematics and Natural Philosophy having been created at Harvard College as early as 1727, a professorship of

Botany in Columbia College in 1792, and a chair of Chemistry at Princeton in 1795. Instruction had also been given in physics and chemistry at the University of Pennsylvania and Dartmouth College, and in physics at Union College. This short list, however, includes all the colleges which had given the physical sciences more than an insignificant place in their curriculums. Even in these the instruction was given by lectures, supplemented at times by experiments which the teachers performed; and anything approaching laboratory work by the student was almost wholly unknown. When Prof. Silliman was elected, in 1801, to the chair of Chemistry, Geology and Mineralogy at Yale College, he visited Dr. Maclean, who was professor of Chemistry at Princeton, and then for the first time saw experiments in chemistry performed.*) Considering the state of scientific knowledge at this period and the general lack of opportunity for the study of science even in Europe, it is not remarkable that this should have been the case in a new country the total population of which in the year 1800 scarcely exceeded that of the city of London to-day.

With the general awakening to the value of the natural sciences, during the first quarter of the century, came provision for their study in other of the academic schools of the country. Within that time courses in various branches were inaugurated at **Yale,** Williams, Bowdoin, Dickinson, William and

\ * Education in the United States, Richard G. Boone.)

Mary, and Hobart Colleges, and in the universities of Georgia, North Carolina and South Carolina. Facilities for practical work by the students were still wanting in nearly all of them, though the apparatus used for illustration had grown in quantity and variety. A chemical laboratory, already mentioned, was in existence at Princeton, one was fitted up at Williams College in 1812, and one at Harvard shortly after this date. A few others were also to be found. They were all, of course, crude and unpretending compared with those thickly scattered over the country to-day. Nor were the steps taken in the study of science always forward. Thus there was organized in the University of Pennsylvania, in 1816, a department of Natural Science "with five professors; and annual courses of lectures, to be publicly delivered, were required by the regulations. The courses of instruction embraced natural philosophy, botany, natural history, mineralogy, chemistry applied to agriculture and the arts, and comparative anatomy. The support given by the public, however, was not sufficient to compensate for the efforts put forth, the professors were badly paid and the department soon fell into neglect. It was abolished shortly after the establishment of the Franklin Institute, in 1824, which rendered, it was said at the time, such a department in the university 'unnecessary.' " *

The time had now come, not only for the addition

---

* Historical Sketch of the University of Pennsylvania, John L. Stewart. Circular No. 2, 1892, of the U. S. Bureau of Education.

of scientific courses to the curriculums of the institutions of learning, but for a general diffusion of scientific knowledge among those who could not have the
advantage of an education higher than that afforded
by the common schools. Attempts in this direction
had already been made in Europe. (When Count
Rumford returned from Munich to London in 1795
he endeavored to interest the people of England, as
he had those of Germany, in his plans for public and
domestic economy, more particularly in the economi-
cal consumption of coal, improvements in the construction of fireplaces and the heating of buildings
by steam. In 1799 he issued in London a prospectus entitled " Proposals for forming by subscription,
in the metropolis of the British empire, a public
institution for diffusing the knowledge and facilitat-
ing the general introduction of useful mechanical
inventions and improvements, and for teaching, by
courses of philosophical lectures and experiments,
the application of science to the common purposes
of life." The result was the establishment, in the
year 1800, of the Royal Institution of Great Britain,
which had for its object the purposes outlined in his
prospectus.)

Other men had not been blind to the benefits
which would accrue to civilization if the people generally could be instructed in the application of science
to the common purposes of life. Franklin's opinions
upon this subject are well known. John Adams
believed that the state should make provision for this

purpose, as is shown by the following extract from the constitution of Massachusetts, of 1780, of which he was the principal author: "to encourage private societies and public institutions, rewards and immunities for the promotion of agriculture, arts, sciences, commerce, trades, manufactures, and a natural history of the country." Jefferson also proposed a school of technical philosophy, to be maintained wholly at public expense, where certain of the higher branches should be taught in abridged form to meet practical wants. "To such a school", he wrote, "will come the mariner, carpenter, shipwright, pump-maker, clock-maker, machinist, optician, metallurgist, founder, cutler, druggist, brewer, vintner, distiller, dyer, painter, bleacher, soap-maker, tanner, powder-maker, salt-maker, glass-maker, to learn, as much as shall be necessary to pursue their art understandingly, of the sciences of geometry, mechanics, statics, hydrostatics, hydraulics, hydrodynamics, navigation, astronomy, geography, optics, pneumatics, acoustics, physics, chemistry, natural history, botany, mineralogy and pharmacy." *

The influence of such opinions gave impetus to the diffusion of scientific knowledge among the people on this continent as well as abroad. Partly for this purpose the Franklin Institute at Philadelphia was founded at the end of the first quarter of the

* Early History of the University of Virginia, as contained in the letters of Thomas Jefferson and Joseph C. Cabell. Edited by J. W. Randolph, Richmond, 1856.

century, and the example of Count Rumford is believed to have been at least one of the causes of the
foundation of another institution which has done
more for science and engineering in this country than
any other school. Although, as before shown, opportunities had been offered in various colleges and
universities for the study of natural science, and the
above-mentioned institutions for popular lectures on
its various branches had been founded here and in
England, there had not been in existence in either
country a school created avowedly for purposes of
scientific instruction ; and there was left to Stephen
Van Rensselaer of Albany, N. Y., the honor of
establishing, at his own expense, an institution with
this as its main object.

It was called the Rensselaer school. (That the
founder had definite ideas not only in relation to the
purposes of the institution but also in regard to its
general management and the methods of instruction
to be pursued, is attested by a letter dated November
5, 1824, to the Rev. Samuel Blatchford of Lansingburgh. It forms the first official notice of the foundation, and reads as follows :

"Dear Sir : I have established a school at the
north end of Troy, in Rensselaer county, in the
building usually called the Old Bank Place, for the
purpose of instructing persons, who may choose to
apply themselves, in the *application of science to the
common purposes of life.* My principal object is, to
qualify teachers for instructing the sons and daughters

OLD BANK PLACE. OCCUPIED, 1824–34 AND 1841–44.

of farmers and mechanics, by lectures or otherwise, in the application of experimental chemistry, philosophy and natural history, to agriculture, domestic economy, the arts and manufactures. From the trials which have been made by persons in my employment at Utica, Whitesborough, Rome, Auburn and Geneva during the last summer, I am inclined to believe that competent instructors may be produced in the school at Troy, who will be highly useful to the community in the diffusion of a very useful kind of knowledge, with its application to the business of living. Apparatus for the necessary experiments has been so much simplified, and specimens in natural history have become subjects of such easy attainment, that but a small sum is now required as an outfit for an instructor in the proposed branch of science; consequently every school district may have the benefit of such a course of instruction about once in two or three years, as soon as we can furnish a sufficient number of teachers. I prefer this plan to the endowment of a single public institution, for the resort of those only whose parents are able and willing to send their children from home or to enter them for several years upon the Fellenberg plan. It seems to comport better with the habits of our citizens and the genius of our government to place the advantages of useful improvement equally within the reach of all.

"Whether my expectations will ever be realized or not. I am willing to hazard the necessary expense of

making the trial. Having procured a suitable building advantageously located among farmers and
mechanics, and having furnished funds which are
deemed sufficient by my agent in this undertaking
for procuring the necessary apparatus, etc., it now
remains to establish a system of organization adapted
to the object. You will excuse me if I attach too
much consequence to the undertaking. But it appears to me that a board of trustees to decide upon
the manner of granting certificates of qualifications,
to regulate the government of students, etc., is essential. I therefore take the liberty to appoint you a
member and president of a board of trustees for this
purpose. I appoint the following gentlemen trustees
of the same board. The Rev. Dr. Blatchford and
Elias Parmalee of Lansingburgh; Guert Van
Schoonhoven and John Cramer, Esqs., of Waterford ;
Simmeon De Witt and T. Romeyn Beck of Albany ;
John D. Dickinson and Jedediah Tracy of Troy.
And I appoint O. L. Holley, Esq., of Troy, and T.
R. Beck of Albany, first and second vice-presidents
of said board.

" As a few regulations are immediately necessary
in order to present the school to the public, it seems
necessary that I should make the following orders,
subject to be altered by the trustees after the end of
the first term.

" *Order* 1. The board of trustees is to meet at
times and places to be notified by the president, or
by one of the vice-presidents, in the absence or disa-

bility of the president. One half of the members of the board are to form a quorum for doing business. A majority of the members present may fill any vacancy which happens in the board ; so that there may be two members resident in Troy, two in Lansingburg, two in Waterford, and two in Albany. The powers and duties of the trustees to be such as those exercised by all similar boards, the object of the school being always kept in view.

" *Order* 2. I appoint Dr. Moses Hale of Troy, secretary, and Mr. H. N. Lockwood, treasurer.

" *Order* 3. I appoint Amos Eaton of Troy, professor of chemistry and experimental philosophy, and lecturer on geology, land surveying, and the laws regulating town officers and jurors. This office to be denominated the senior professorship.

" *Order* 4. I appoint Lewis C. Beck of Albany, professor of mineralogy, botany and zoology, and lecturer on the social duties peculiar to farmers and mechanics. This office to be denominated the junior professorship.

" *Order* 5. The first term is to commence on the first Monday in January next, and to continue fifteen weeks. For admission to the course, including the use of the library and reading-room, each student must pay twenty-five dollars to the treasurer, or give him satisfactory assurances that it will be paid in one year. In addition to this, each section of students must pay for the chemical substances they consume and the damage they do to apparatus.

"*Order* 6. All the pay thus received by the treasurer, as for parts of courses of instruction, is to be paid over to said professors as the reward of their services.

/ "*Order* 7. In giving the course in chemistry the students are to be divided into sections, not exceed-ing five in each section. These are not to be taught by seeing experiments and hearing lectures, accord-ing to the usual method. But they are to lecture and experiment by turns, under the immediate direc-tion of a professor or a competent assistant. Thus by a term of labor, like apprentices to a trade, they are to become operative chemists.

"*Order* 8. At the close of the term each student is to give sufficient tests of his skill and science be-fore examiners, to be appointed by myself, or by the trustees if I do not appoint. The examination is not to be conducted by question and answer; but the qualifications of students are to be estimated by the facility with which they perform experiments and give the rationale; and certificates or diplomas are to be awarded accordingly. |

"*Order* 9. One librarian, or more, to be ap-pointed by the professors, will be keeper of the read-ing-room. All who attend at the reading-room are to respect and obey the orders of the librarian in regard to the library and conduct while in the room.

"*Order* 10. Any student who shall be guilty of disorderly or ungentlemanly conduct is to be tried and punished by the president or vice-president and two trustees. The punishment may extend to expulsion and forfeiture of the school privileges, without a release from the payment of fees. But a student may appeal from such decision to the board of trustees.

"This instrument, or a copy of it, is to be read to each student before he becomes a member of the school; and he is to be made to understand that his matriculation is to be considered as an assent to these regulations.

"STEPHEN VAN RENSSELAER.

"ALBANY, Nov. 5, 1824."

This document shows the aim of the founder of the Rensselaer School to have been substantially that of the originator of the Royal Institution, though the methods pursued in attaining the object sought were different. He was doubtless familiar with the work and writings of Rumford, and it will be noticed that he has used in his description of the purpose of the school the same expression found in the London prospectus of 1799—"the application of science to the common purposes of life." * Attention will be given later to the peculiar methods of instruction outlined in this letter, and before proceeding with the history of the school a short account will be given of the lives of its founder and of another to whose talent as a teacher and scientific investigator the success of the school was largely due.

* See the address of President James Forsyth in Proceedings of the Semi-Centennial Celebration of the Rensselaer Polytechnic Institute, 1874.

CHAPTER II.

STEPHEN VAN RENSSELAER was the fifth in direct
line of descent from Killian Van Rensselaer, a merchant of Holland, who obtained by purchase from
the Indians, about the year 1637, a district about
twenty-four miles in breadth by forty-eight in length,
comprising the territory which has since become the
counties of Albany, Columbia and Rensselaer, in the
state of New York. He named it the Colony and
Manor of Rensselaerwyck, and was its first Patroon.
Stephen was born November 1, 1764, in the city
of New York. His father was Stephen Van Rensselaer, the seventh Proprietor or Patroon of Rensselaerwyck, and his mother was Catharine, the daughter of Philip Livingston. Upon the death of his
father in 1769, the care of the estate, which fell
exclusively to him by the law of primogeniture, de-
volved upon his uncle, General Ten Broeck, who
also acted as guardian during his minority. He was
at first sent to a school in Albany and afterwards to
one in Elizabethtown, New Jersey. At the beginning of the Revolution he was removed to Kingston,
N. Y., and acquired the elements of a classical education at the Kingston. Academy. He was later

sent to Princeton College, but, in consequence of its proximity to the seat of war, it was thought advisable to send him to Harvard College, where he was graduated as a Bachelor of Arts in 1782, in the nineteenth year of his age. Returning to Albany he married, in 1783, a daughter of General Philip Schuyler, and upon reaching his majority settled down in the Manor House and took charge of his estates. By offering leases for long terms at a very moderate rent he succeeded in bringing a large portion of his land into cultivation, but little of which had, until then, been converted into farms, and thus secured for himself a competent income.

He was made a major of infantry in 1786, and when, in 1801, Governor Jay formed the cavalry of the state into a separate corps he was placed in command with a commission of major-general of cavalry. He was elected, as a federalist, to the Assembly of the State in 1789, and the next year became a state senator, which position he held until 1795, when he was chosen lieutenant-governor at the same time that John Jay was elected governor. He was lieutenant-governor for six years, and was nominated for governor in 1801, but was defeated by De Witt Clinton. In the same year he was a member of the constitutional convention, and presided over it during the greater part of its deliberations. He was again elected to the Assembly in 1807, and when, during this term, a project was agitated to appoint a commission for exploring a route for a western canal, he was strongly in favor of it. Having been appointed, in 1810, to serve on this commission, he, in company with the other members, made an exploration of the route for a canal from the Hudson River to Lake Erie.

When war with great Britain was declared in 1812, he was given the command of the state militia, and on the 13th of October of that year assaulted and took the Heights of Queenstown, Canada, from which, however, he was compelled to withdraw by the refusal of the state militia, under the plea of constitutional scruples, to leave the state. His services in the field ended with this campaign, and in 1813 he was again nominated for governor, but was defeated by a small majority. In the meantime the canal commission had continued its existence, and in 1816, when the Legislature directed the construction of the Erie Canal and committed the execution of the work to a board of canal commissioners, he was made a member of that body, and was its president from April, 1824, until his death. He was again elected a member of Assembly in 1816, in 1819 became a Regent of the State University, of which he was chancellor from 1835 until his death, and was a member of the constitutional convention of 1821.

From his position as Patroon and because of the great extent of territory he possessed, as well as on account of his great intelligence and the benevolence of his nature, Stephen Van Rensselaer had always

been strongly in favor of the encouragement of farmers and the improvement of agriculture. When,
therefore, in 1819, an act for the encouragement of
agriculture was passed by the Legislature of the
State, under the provisions of which delegates from
county societies formed a Central Board of Agriculture, he was elected its president at the first meeting in Albany, in January, 1820. Although the life
of the board was brief, it was long enough to permit
a geological and agricultural survey of the counties
of Albany and Rensselaer to be made under its
direction, though at the expense of its president.
This survey was executed by Professor Amos Eaton
with the aid of two assistants, and was the first attempt made in this country to collect and arrange
geological facts with a direct view to the improvement of agriculture. Analyses of soils were included, as well as a consideration of the proper
methods of culture adapted to them, and the results
were published in three volumes of Transactions and
Memoirs. Imbued with strong opinions as to the
value of such scientific investigations, when the
board ceased to exist Stephen Van Rensselaer was
unwilling to discontinue work of this character, and
in the years 1822 and 1823 he caused to be made,
at his own expense, under the direction of Professor
Eaton, a geological survey extending from Boston to
Lake Erie, a distance of about five hundred and fifty
miles. It embraced a belt fifty miles in width, which
covered, in this State, the line of the Erie canal.

The intelligence and benevolence of the subject of this sketch were now, when he had reached the age of sixty years, to be directed into a new channel. He had long been interested in the instruction of the poorer families of his tenantry, and had reached the conclusion that the most valuable education to be given the masses engaged in the ordinary occupations of life was one which would enable them to apply the principles of science to the " business of living". His first step in this direction was to secure the services of Professor Eaton, with whose qualifications he was thoroughly familiar. He employed him, in the summer of 1824, to traverse the State on or near the line of the Erie canal, provided with sufficient apparatus and specimens to deliver, in all the principal towns where an audience of business men or others could be collected, a series of lectures, accompanied with experiments and illustrations, on " chemistry, natural philosophy and some or all the branches of natural history." This undertaking was entirely successful. Encouraged by it, he determined to establish an institution one of the principal objects of which should be "to qualify teachers for instructing the sons and daughters of farmers and mechanics, by lectures or otherwise, in the application of experimental chemistry, philosophy and natural history to agriculture, domestic economy, the arts and manufactures"; and there resuited the foundation at Troy, N. Y., in 1824, of the school which is the subject of this historical sketch. He at first intended to sustain the school for three years only, expecting that, if at the end of this period it were successful, the public would maintain it. Besides the expense of its original establishment he bore, however, until his death fourteen years later, about one half the cost of its maintenance. As will be seen hereafter, the course of instruction was considerably enlarged, during his life and with his approval, to meet the growing demand for educated engineers and scientific men.

In the meantime, in 1823, General Van Rensselaer had been elected to Congress as a Representative from Albany county, and some of his instructions in relation to the new school were forwarded from Washington. He continued in Congress for six years, and was during this period chairman of the Committee on Agriculture. During a part of his active public life, from 1793 until his resignation in 1819, he was a trustee of Williams College. In 1825 the degree of LL.D. was conferred on him by Yale College. He died at the old Manor House in Albany on the twenty-sixth day of January, 1839.*

Although distinguished because of his position and character, and on account of many years of successful public service in important positions, the memory of Stephen Van Rensselaer will be per-

---

* See " A Discourse on the Life, Services and Character of Stephen Van Rensselaer ", by Daniel G. Barnard, Albany, 1839.

petuated chiefly by means of the school which he established for the benefit of his fellow-men.

In an article on the Institute, one of an interesting series on the engineering schools of the United States, written in 1892 for *Engineering News* by A. M. Wellington, he says : " The founder was not of the class of rich men who found colleges only from a vague philanthropic instinct and to perpetuate his name. He had distinct and very original and decided views as to proper methods of instruction, which he took great pains to provide for and enforce at length. His love of thoroughness, his determination that the instruction should be of the best, if there was any, and that the school should take a high rank among the kindred institutions of the world, crop out constantly in his letters and deed of foundation. . . He was no common founder, and he founded no common school. The cause of engineering education owed much to him indeed."

It will be noticed in the account just given of his life that in all his efforts for the advancement of scientific knowledge, whether by agricultural and geological surveys or by the more direct method of instruction, he employed one individual as his agent. That no error was made in the choice is proved by the uniform success of his endeavors.

Amos Eaton was indeed no ordinary man. The history of the last seventeen years of his life is identical with that of the Rensselaer Institute. The importance of his work, however, not only in the early

Amos Eaton

development of the school but as a scientific investigator and author of works on the natural sciences,
renders it advisable to give, in this connection, a
sketch of his earlier history. He was a native of
Chatham, N. Y., and was born May 17, 1776. His
father, Abel Eaton, was a farmer in comfortable circumstances. He early manifested superior abilities,
and was selected to deliver an oration on the Fourth
of July, 1790, when but fourteen years of age.
About this time, having acted as chainman during a
land survey, he determined to become a surveyor.
Not having the requisite instruments, he interested a
skilful blacksmith in his behalf, who agreed to work
for him at night if he would " blow and strike " by
day. A needle and a good working chain were the
result of several weeks' work. This circumstance in
his life doubtless gave rise to the remark, found in
Silliman's Journal, that " in 1791 he was an ap-
prenticed blacksmith." The bottom of an old pewter
plate, well smoothed, polished and graduated, served
as a compass-circle, so that Eaton, when sixteen
years old, was in the field with his home-made instruments, doing occasional surveying in the neigh-
borhood. He aspired, however, to higher attainments, and, encouraged by his parents, was fitted for
college at Spencertown, N. Y., and was graduated
at Williams College, in 1799, with a high reputation
for scientific knowledge. In the same year he began
the study of law at Spencertown, and subsequently
continued his studies in New York.

At this time he first became interested in the study of botany and other natural sciences. While in New York, in 1802, he borrowed Kirwan's "Mineralogy", then a scarce book, and made a manuscript copy of the entire work. He was admitted to the bar, at Albany, in 1802, and soon after established himself as a lawyer and land agent in Catskill, N. Y. Here he remained several years, his position affording him excellent opportunities for cultivating his growing taste for the natural sciences. In May, 1810, he made in Catskill, it is believed, the first attempt in this country at a popular course of lectures on botany, compiling for the use of his class a small elementary treatise. For this Dr. Hosack, who had formerly taught him in New York, complimented him as being the "first in the field".

Having found his love for the details of his profession diminishing and his interest in the natural sciences increasing, he finally resolved to abandon the practice of law and to fit himself more thoroughly for scientific pursuits. With this end in view he went to New Haven, in 1815, to avail himself of the advantages found at Yale College. He placed himself under the instruction of Professor Silliman, who threw open to him his lectures on chemistry, geology and mineralogy, as well as his own library and the cabinet of minerals of that institution. Here, also, he found a good botanist in Dr. Eli Ives, Professor of Botany and Materia Medica in the medical department of the college, who had accumulated a good library, to which he gave Eaton free access. With these advantages and his already advanced acquirements he was soon well qualified as an explorer and teacher. Returning to Williamstown in 1817, he gave courses of lectures in botany, mineralogy and geology to volunteer classes of students. His influence in the college was remarkable, and he awakened there an interest in the natural sciences which has never died out. His pupils published, in 1817, the first edition of his "Manual of Botany", a 12mo of 164 pages, which, as the late Dr. Lewis C. Beck wrote in 1852, "gave an impulse to the study of botany in New England and New York, as the only descriptive work which was then current was that of Pursh, an expensive one with Latin descriptions." This work was improved by repeated revisions and additions, and became, in the eighth edition, published in 1840, a large octavo volume of 625 pages, which was entitled "North American Botany", and contained a description of 5267 species of plants.

The encouragement received by Mr. Eaton at Williams College determined him to give courses of popular scientific lectures, accompanied with practical instructions, to such classes as he might be able to organize in several of the larger towns of New England and New York. These met with great success, and in the course of two or three years he diffused a great amount of scientific knowledge, and there sprang up as the result of his labors an army of young botanists and geologists. According to Professor Albert Hopkins, of Williams College, he was one of the first to popularize science in the Northern States, and was one of the first in this country to study nature in the field, with his classes.

In 1818, in compliance with a special invitation from Governor DeWitt Clinton, he went to Albany and delivered a course of lectures before the members of the Legislature. Here he became acquainted with many of the leading men of the State, interesting them especially in geology and its application, by means of surveys, to agriculture. A train of causes was thus set in operation which resulted in giving to the world that great work, "The Natural History of New York", so creditable to the State and to the scientific men who executed it, of whom several had been Prof. Eaton's pupils. In this year he published the first edition of his " Index to the Geology of the Northern States", which was the first attempt at a general arrangement of the geological strata in North America. In his " Education in the United States", Boone says : " Among the older geologists, and one of the first to study nature in the field, was Prof. Amos Eaton of Williams College. He has been called the ' Father of American Geology ', was the instructor of Hall, Dana and Williams, and initiated the interest in a half dozen states."

He afterwards delivered several courses of lectures in the medical college at Castleton, Vt., in which he was appointed Professor of Natural History in 1820.

In this year and the following one he made the geological and agricultural surveys of Albany and Rensselaer counties to which reference has been made in
the sketch of the life of Stephen Van Rensselaer. Of
these surveys Professor Silliman remarked, in his Journal, " The attempt is novel in this country "; adding,
" We are not aware of any attempt, on so extensive
and systematic a scale, to make them subservient
to the important interests of agriculture." There has
also been previously mentioned the geological survey
of the district adjoining the Erie Canal, made by
Professor Eaton in 1822 and 1823. A report of this
survey, consisting of 160 octavo pages, with a profile
section of rock formations from the Atlantic Ocean,
across the states of Massachusetts and New York, to
Lake Erie, was published in 1824. In relation to
this work Governor Seward, in his introduction to
the " Natural History of the State of New York",
said : " This publication marked an era in the progress of geology in this country. It is in some respects inaccurate, but it must be remembered that its
talented and indefatigable author was without a guide
in exploring the older formations, and that he described rocks which no geologist had, at that time,
attempted to classify. Rocks were then classified
chiefly by their mineralogical characters, and the aid
which the science has since learned to derive from
fossils, in determining the chronology and classifica-
tion of rocks, was scarcely known here and had only
just begun to be appreciated in Europe. We are

indebted, nevertheless, to Prof. Eaton for the commencement of that independence of European clas-
sification which has been found indispensable in
describing the New York system." He also said:
" Prof. Eaton enumerated nearly all the rocks in
western New York, in their order of succession; and
his enumeration has, with one or two exceptions,
proved correct. It is a matter of surprise that he
recognized, at so early a period, the old red sand-
stone on the Catskill mountains, a discovery the
reality of which has since been proved by fossil
tests."

Such was the man chosen by Stephen Van Rensselaer to take charge, as Senior Professor and Agent,
of the institution which he established in 1824.
Eaton's enthusiasm and remarkable powers as a
teacher doubtless had their influence in determining
him to bear the expense of the series of lectures in
towns along the Erie Canal, and afterwards to under-
take the creation of the school. And it does not
detract from the credit of the founder to say that the
methods and the object of the institution, as set forth
in his letter to Dr. Blatchford, were, if not wholly, at
least partly, due to its first Senior Professor.

Rev. Calvin Durfee in his History of Williams
College (1860), from which most of this account of
the life of Eaton is taken, says: " In this school
Prof. Eaton was able to perfect and carry out, to a
high degree of success, his favorite plan of teaching
classes by making his pupils experimenters and

MAIN BUILDING. COMPLETED, 1864.

workers in every department of science where it was practicable ; substituting also lectures by the pupils to each other in place of the usual system of recitations. This method of giving instruction and of preparing young men to become successful teachers has here succeeded most admirably, and has been, in some of its features, introduced into other schools of science." And again : "The history of natural science on this continent can never be faithfully written, without giving the name of Amos Eaton an honorable place. It was he, more than any other individual in the United States, who, finding the natural sciences in the hands of the learned few, by means of his popular lectures, simplified text-books and practical instructions, threw them broadcast to the many. He aimed at a general diffusion of the natural sciences, and nobly and successfully did he accomplish his mission."

The last seventeen years of his life were passed in Troy as Senior Professor in the Rensselaer School or Rensselaer Institute, the name by which it was afterwards known. In the minutes of the Board of Trustees we find this tribute to his memory : "The trustees are called to the painful duty of recording the death of Prof. Amos Eaton, who has long been at the head of the Rensselaer Institute. He died on the tenth day of May, 1842, in the sixty-sixth year of his age. It is but simple justice to say that Prof. Eaton was, under its distinguished patron and benefactor, the founder of this school of the

natural sciences; that he was a faithful and successful instructor in these studies, and that he contributed,
by his labors in the Institute and by his geological
survey of the State of New York, more than any
other man in our country to the cultivation of geological science.   While the trustees consider the experiment, as to the mode of communicating knowledge adopted in the Rensselaer Institute, as a successful one, they are fully persuaded that much of
this success is due to the industry and enthusiasm of
Prof. Eaton.   Few men were ever more devoted to
the peculiar duties of his profession than he, and his
perseverance was equal to his devotedness.   His
removal may be considered not only as a loss to our
city but to our country."

An idea of his labors as an author and investi-
gator may be obtained from a list of his works.   He
published an Elementary Treatise on Botany, 1810 ;
Manual of Botany, 1817 ; Botanical Dictionary, 1817 ;
Botanical Exercises, 1820 ; Botanical Grammar and
Dictionary, 1828; Chemical Note Book, 1821 ;
Chemical Instructor, 1822 ; Zoölogical Syllabus and
Note Book, 1822 ; Cuvier's Grand Division, 1822 ;
Art Without Science, 1800 ; Philosophical Instructor,
1824; Directions for Surveying and Engineering,
1838 ; Index to the Geology of the Northern States,
1818 ; Geological and Agricultural Survey of the
County of Albany, N. Y., 1820; Geological and
Agricultural Survey of Rensselaer County, 1822 ;
Geological Nomenclature of North America, 1822 ;

Geological and Agricultural Survey of the District adjoining the Erie Canal, 1824; Geological Text Book, prepared for popular lectures on North American geology, 1830; Geological Note Book for Troy Class, 1841. Of most of these works a number of different editions were published.

In after years his memory as a botanist was honored by Professor Gray, who named for him two species of plants, the *Eatonia obtusata* and *Eatonia Pennsylvanica.*

Enough has been said to show the great value of his original work in the natural sciences, and this short sketch of his life will be closed by a tribute to his memory as a teacher, paid, thirty years after his death, by one of his former pupils. At the ceremonies attending the erection of a monument to him, during the celebration of the semi-centennial of the Rensselaer Polytechnic Institute, in 1874, Professor James Hall, of the class of 1832, New York State Geologist and Palæontologist, himself with a world-wide fame in his specialty, said, in part: " In the progress of civilization, 'it is not the slow uniform motion of the great masses that helps it forward, but the few men who come out from them and strike a new key. Prof. Eaton taught us the manipulations in science with the simplest materials, so that a student could go into the forest and construct a pneumatic trough, or a balance, and perform there his experiments in chemistry or physics. To his memory we owe much. His name has been neglected before the public, but cherished in the bosoms of those who knew him—a man capable of interesting young men, having a brain one fourth larger than that of the mass of mankind, and that brain devoted to the service of science. If we with great means do what he did with small, we shall deserve well of coming generations."

CHAPTER III.

SHORTLY after the receipt of Stephen Van Rensselaer's letter, given in the first chapter, the Rev. Dr.
Blatchford called together the Board of Trustees of
the new school. The first meeting was held December 29, 1824, and the institution was then named the
" Rensselaer School". An outline of the method of
instruction to be pursued may be gathered from the
minutes of the proceedings of this meeting, during
which it was :

/" *Resolved*, That persons attending the courses of
instruction at Rensselaer School be distributed into
three classes, viz.: a Day Class, an Afternoon Class
and an Evening Class.\

" The exercises of the Day Class, for six hours in
each day, except Sunday, shall consist of experiments
in chemistry, performed by themselves and in giving
explanations, or the *rationale* of the experiments ;
and they shall undergo daily examinations and alter-
nately become examiners themselves. Each member of this class shall pay $25 a term (as prescribed
by the founder in the orders promulgated by him),
and at the end of each term shall be examined for
his certificate.

"The Afternoon Class shall consist of those who may have previously attended one or more courses of lectures on chemistry at some public institution. They will hear no afternoon lectures; but their exercises will consist of a course of experiments in chemistry, performed by themselves, as above, with the *rationale*, conducted under the superintendence of the senior professor. These exercises will occupy three hours in the afternoon of each week-day except Saturday. Each member of this class shall pay $10 a term, and at the end of each term undergo an examination for his certificate.

"The Evening Class will attend lectures, on three evenings of each week, for ten weeks. This course of lectures will embrace chemistry, experimental philosophy and the outlines of mineralogy, geology, botany and zoology. The charge for attendance will be $5. Members of this class will not be examined at the end of the term, but may have certificates of attendance."* |

The opening of the school on Monday, January 3, 1825, was announced by a notice, signed by the president, printed in the Troy *Sentinel* of December 28th. The announcement reads, in part, as follows:

"The Hon. Stephen Van Rensselaer having established a school near the northern limits of Troy for teaching the physical sciences with their application to the arts of life; having appointed Profs. A. Eaton and L. C. Beck to give courses of instruction particu-

* Troy *Sentinel*, January 4, 1825.

CHEMICAL LABORATORY. COMPLETED, 1866.

larly calculated to prepare operative chemists and practical naturalists, properly qualified to act as teachers in villages and school districts ; having appointed an agent and furnished him with funds for procuring apparatus and fitting up a laboratory, library-room, etc. ; and the agent having given notice to the president of the institution that the requisite collections and preparations are completed, it seems proper to give public notice of these circumstances.

" Accordingly the public is respectfully notified that everything is in readiness at the Rensselaer School for giving instruction in chemistry, experimental philosophy and natural history, with their application to agriculture, domestic economy and the arts ; and also for teaching land surveying. . .

" During the day no lectures will be given by the professors, but under their superintendence the students, divided into sections, will perform all the experiments and give the explanations, the students thus acting as lecturers and the professors as auditors. . .

" Students who wish for *extra* accommodations will pay from $1.75 to $2.00 a week for board and lodging. But any number of students can have good plain board and lodging near the school for $1.50 a week."

The courses and methods thus set forth are seen to be those outlined in the letter of the founder, with the orders accompanying it ; and the trustees, instructors and other officers were the persons named by him in the same document. Being at this time a member of Congress, Mr. Van Rensselaer wrote from Washington another letter to Dr. Blatchford, dated February 11, 1825, in which a draft of by-laws for the further government of the institution was enclosed. As these two letters are both important documents in the early history of the school, being adopted as its constitution at the first meeting of the trustees and the two following ones, held March 11 and June 5, 1825, the second letter will also be given in full :

"WASHINGTON, February 11, 1825.

"*Dear Sir:* I offer my acknowledgements for the interest you have taken in promoting the school over which you preside. I have enclosed a draft, hastily drawn up, of by-laws, for the government of the school, which I beg to submit to yourself and the gentlemen associated with you for consideration and amendment. I flatter myself that the school will succeed and that the advantages I anticipated will be realized.

"With respect, yours sincerely,

"S. V. RENSSELAER."

[ENCLOSED DRAFT.]

" 1. That there be two terms in each year, of twelve or fifteen weeks each, to be called the summer term and winter. The summer term to commence in May, the winter term to commence in January—say, the last of May and January.

" 2. That during the summer term the students shall be taught the elementary principles of the science of chemistry, experimental philosophy, natural history, land surveying, etc., with their application to agriculture, manufactures and the arts.

" 3. That, with the consent of the proprietors, a number of well cultivated farms and workshops in the vicinity of the school be entered on the records of the school as places of scholastic exercise for students, where the application of the sciences may be most conveniently taught.

" 4. That during the winter term students be exercised in giving lectures, by turns, on all the branches taught in the summer term, under the direction of the professors or their assistants, in order to qualify them for giving instruction in these branches. And that a course of evening lectures be given in the winter term, by the professors, so as to embrace elementary views of the whole course of instruction given at the school.

" 5. That an annual commencement be held in April at the close of the winter term, for conferring diplomas on those qualified."

After about fourteen months of successful trial the school was incorporated by the following act, passed March 21, 1826:

AN ACT TO INCORPORATE THE RENSSELAER SCHOOL.*

Whereas, the honorable Stephen Van Rensselaer has procured suitable buildings in the city of Troy, in Rensselaer

* Laws of the State of New York, 1826, Chap. 83.

county, and therein set up a school, and at his own private expense has furnished the same with a scientific library, chemical and philosophical; instruments for teaching land surveying and other branches of practical mathematics, which are useful to the agriculturist, the machinist, and to other artists, has caused to be prepared and furnished separate and commodious rooms for instruction in natural philosophy, natural history, the common operations in chemistry, and an assay-room for the analysis of soils, manures, mineral and animal and vegetable matter, with the application of these departments of science to agriculture, domestic economy, and the arts : *And whereas,* said Van Rensselaer has employed teachers, and caused an experimental system of instruction to be adopted by them, whereby each student is required to observe the operations of a select number of agriculturists and artists in the vicinity of said school, and to demonstate the principles upon which the results of such operations depend, by experiments and specimens performed and exhibited by his own hands, under the direction of said teachers : *And whereas,* one important object of said school is to qualify teachers for instructing youths in villages and in common-school districts, belonging to the class of farmers and mechanics, by lectures or otherwise, in the application of the most important principles of experimental chemistry, natural philosphy, natural history and practical mathematics to agriculture, domestic economy, the arts and manufactures : *And whereas,* the trustees of said school, who were appointed to take charge thereof by said Van Rensselaer, by an instrument in writing dated November the fifth, in the year eighteen hundred and twenty-four, have represented to this Legislature, that after having tested the plan of said school by a trial of one year, they find it to be practicable and in their opinion highly beneficial to the public : *And whereas,* the legislature considers it to be their duty to encourage such laudable efforts and such munificent applications of surplus wealth of individuals : Therefore

1. *BE it enacted by the People of the State of New York, represented in Senate and Assembly,* That Simeon De Witt,

Samuel Blatchford, John D. Dickinson, Guert Van Schoonhoven, Elias Parmalee, Richard P. Hart, John Cramer and Theodore Romeyn Beck, shall be and hereby are constituted a body corporate and politic, by the name of "the president and trustees of Rensselaer School," and by that name they shall have perpetual succession, and shall be capable of suing and being sued, pleading and being impleaded, answering and being answered unto, defending and being defended, in all courts and suits whatsoever ; and may have a common seal, with power to change or alter the same from time to time, and shall be capable of purchasing, taking possession of, holding and enjoying to them and their successors any real estate, in fee simple or otherwise, and any goods, chattels, and personal estate, and of selling, leasing, or otherwise disposing of the said real and personal estate, or of any part thereof, at their will and pleasure ; *Provided, however,* That the funds of said corporation shall be used for and appropriated to the objects contemplated in the preamble of this act ; *And provided also,* That the clear annual income of such real and personal estate shall not exceed the sum of twenty thousand dollars.

2. *And be it further enacted,* That the said trustees shall, from time to time, forever hereafter have power to make, constitute, ordain and establish such by-laws and regulations as they shall judge proper, for the election of the officers and prescribing their respective functions, for the government of the officers and students of said school as to their respective duties, for collecting fines, impositions, and term fees, for suspending, expelling, and otherwise punishing students, so that it shall not extend further than expulsion and retaining term fees, and collecting the amount of any damage done by students to the property of said school ; for conferring on students such honors as they may judge proper, having relation to the object of said school as expressed in the said preamble, and for managing and directing all the concerns of said school; also for confirming the constitution and by-laws, or any part thereof heretofore adopted by said trustees, provided such by-laws and regulations have relation to the subjects of the preamble of this act exclusively.

3. *And be it further enacted,* That the officers of said school shall consist of a president, two vice-presidents, a treasurer and secretary, two professors, and such a number of adjunct professors and assistants as the trustees may from time to appoint or authorize the appointment of, a librarian, monitor and steward.   That whenever any vacancy shall happen among the trustees of said school, such vacancy or vacancies may be filled by a quorum of the remaining trustees, so that two trustees shall reside in Albany, two in Troy, two in Lansingburgh, and two in Waterford.

4. *And be it further enacted,* That there shall be one annual meeting of the trustees of said school on the last Wednesday in April, at which meeting four members of the board of trustees shall constitute a quorum, and that four members shall also constitute a quorum at all special meetings, to be called by the president at any time after the passing of this act, provided a written notice of such meeting, signed by the president or by one of the vice-presidents, shall be left at the dwelling-house or place of residence of such member of the board seven days previous to such special meeting.

5. *And be it further enacted,* That Samuel Blatchford shall be president, and that he, together with all the other officers of the said school, shall remain as heretofore, until a special meeting of a quorum of said trustees shall be assembled at such school, by the president, or by a vice-president, as prescribed in the fourth section of this act, or until the annual meeting on the last Wednesday in April next, then to be permitted to continue in their respective offices, or their places to be filled at the pleasure of the trustees.

6. *And be it further enacted,* That the legislature may at any time modify or repeal this act.

Upon the passage of the act of incorporation the trustees named therein held a meeting at the school on April 3, 1826, and, after reappointing all the officers who had been serving at the time the bill was passed, they resolved that the constitution pre-

RANKEN HOUSE.   PURCHASED, 1877.

viously adopted, consisting of the two letters of Mr. Van Rensselaer, should continue to be the constitution of the school, with certain amendments. These amendments provided that there should be three terms in each year, to be called the Fall Term, Winter Term and Spring Term; that the fall term should be an experimental term commencing on the third Wednesday in July and continuing fifteen weeks; that the winter term should be a recitation term commencing on the third Wednesday in November and continuing twelve weeks; that the spring term should be an experimental term commencing on the first Wednesday in March and continuing until the last Wednesday in June, and that the last mentioned day should be the Annual Commencement.

At the same meeting a code of by-laws consisting of eleven articles was passed. Some of these articles which embody the curriculum of that day will be given in full.

" Article 1. The course of exercise at said school in the Fall Term shall be, as nearly as circumstances will permit, as follows : Each student shall give five lectures each week on systematic botany, demonstrated with specimens, for the first three weeks, and shall either collect, analyze and preserve specimens of plants, or examine the operations of artists and manufacturers at the school workshops, under the direction of a professor or assistant, who shall explain the scientific principles upon which such operations depend, four hours on each of six days in every week, unless excused by a professor on account of the weather, ill health or other sufficient cause. For the remaining twelve weeks, each student shall give fifteen lectures on mineralogy and zoology, demonstrated with specimens; fifteen lectures on chemical powers and substances not metallic; fifteen lectures on natural philosophy, including astronomy; and fifteen lectures on metalloids, metals, soils, manures, mineral waters, and animal and vegetable matter—all to be fully illustrated with experiments performed with his own hands; and shall examine the operations of artists at the school workshops, under the direction of a professor or assistant, four hours on every Saturday, unless excused as aforesaid.

"Article 2. During the Winter Term students shall recite, to a professor or to a competent assistant, the elements of the sciences taught in the fall and spring terms; and shall study and recite, as auxiliary branches in aid of these sciences, rhetoric, logic, geography, and as much mathematics as the faculty shall deem necessary for studying land surveying, common mensuration, and for performing the common astronomical calculations.

"Article 3. The course of exercises in the Spring Term shall be, as nearly as circumstances will admit, as follows : Each student shall, during the first six weeks, give ten lectures on experimental philosophy; ten lectures on chemical powers and on substances not metallic; and ten lectures on metalloids, metals, soils and mineral waters. For the remainder of the

term each student shall be exercised in the applica-
-tion of the sciences before enumerated to the analysis of particular selected specimens of soils, manures,
animal and vegetable substances, ores and mineral
waters ; and shall devote four hours of each day, unless excused by one of the faculty, to the examination of the operations of the agriculturists on the
school farms, together with the progress of cultivated
grains, grasses, fruit-trees and other plants, to practical land surveying and general mensuration, to calculations upon the application of water-power and
steam which is made to the various machines in the
vicinity of the school, and to an examination of the
laws of hydrostatics and hydrodynamics which are
exemplified by the locks, canals, aqueducts and natural waterfalls surrounding the institution."

Article 4 relates to the admission of students. It
provides that no candidate shall be admitted as an
annual student under the age of seventeen years.]
The conditions under which examinations are to be˙
held and degrees given are set forth in Article 5.
The degree conferred was Bachelor of Arts in Rensselaer School, A.B. (r.s.). After the expiration of
three years from the receipt of this degree, or of one
year, if the student attended a second annual course
at the school and proved his capacity, the degree
Master of Arts in Rensselaer School, M.A. (r.s.),
was conferred. No degree could be conferred on
any one less than eighteen years old ; and in using
the abbreviation for Bachelor or Master of Arts

the letters (r.s.) had to be added. It is provided in Article 6 that, after receiving a degree, a person ever after remained a member of the school, and must, every three years, report his occupation to the trustees. We learn from Article 7 that at this time the tuition was $15 for each experimental term and $6 for the recitation term. The student also had to pay extra for breakage and chemicals consumed and his proportion of the cost of fuel and lights and the services of the monitors. Article 8 relates to weekly reports from professors, Article 9 to the times of meeting of the board of trustees, Article 10 makes void all previous rules and by-laws, and Article 11 provides for temporary rules to be made by the faculty.

Much of the information above given in relation to the founding of the school is taken from the original minutes of the meetings held by the board of trustees and from a pamphlet entitled " Constitution and Laws of Rensselaer School in Troy, New York ; adopted by the board of trustees April 3, 1826; together with a Catalogue of Officers and Students", which was published in Albany in 1826. Among " Notices and Remarks" found in it there is a paragraph containing an itemized account of the necessary expenses of a student. This will be quoted to show the difference between the cost of education at that time and the outlay required at the present day : "The expenses for a student of ordinary prudence will be about $100, if he is absent during the winter term :

Board, 30 weeks at $1.50..........$45.00
Washing, about 18 cents per week..   5.62
Chemical substances, etc., about.....   4.00
Proportion of fuel and lights, about..   6.00
Text-books, about................   4.00
Experimental term fees, $15........ 30.00
$$\overline{}$$
Total.....................$94.62"

The catalogue contains the names of the professors and twenty-five students. Amos Eaton is entitled professor of chemistry and natural philosophy and lecturer on geology, land surveying, etc., and Lewis C. Beck, professor of botany, mineralogy and zoology. Eighteen of the students came from the state of New York, two from New Hampshire, two from Massachusetts, one from Vermont, one from Ohio and one from Pennsylvania.

CHAPTER IV.

METHODS OF INSTRUCTION—PREPARATION
BRANCH ESTABLISHED.

ALTHOUGH a general knowledge of the mode of instruction pursued at the Rensselaer School may be obtained from the letters of the founder and especially from the by-laws adopted by the board of trustees, April 3, 1826, the novelty of the system of teaching and the fact that the institution was the first school of science established either in this country or Great Britain, renders advisable a more detailed account of its methods at that time. The peculiarities of the school are described in several of the earlier pamphlets published under the auspices of the board of trustees. Its three distinct characteristics will be given in the words of one of these publications.

"1. The most distinctive character in the plan of the school consists in giving the pupil the place of teacher in all his exercises. From schools or colleges where the highest branches are taught to the common village schools, the teacher always improves *himself* more than he does his *pupils*. Being under the necessity of relying upon his own resources and of making every subject his own, he becomes an

WILLIAMS PROUDFIT ASTRONOMICAL OBSERVATORY. COMPLETED, 1878.

adept as a matter of necessity. Taking advantage of this principle, students of Rensselaer School learn by giving experimental and demonstrative lectures, with experiments and specimens.

" 2. In every branch of learning the student begins with its practical application, and is introduced to a knowledge of elementary principles, from time to time, as his progress requires. After visiting a bleaching-factory he returns to the laboratory and produces the chlorine gas and experiments upon it until he is familiar with all the elementary principles appertaining to that curious substance. After seeing the process of tanning he enters the laboratory with most ardent zeal for a knowledge of the principles upon which the tanner's operations depend. He can now apply the experiment for making an insoluble precipitate of tannin and animal gelatin, also the soapy compound of animal oil and an alkaline earth, etc. After seeing buhr millstones consolidated by a gypsum cement, he is anxious to try the experiment of disengaging the water of combination in the gypsum, to observe the effect of reabsorption. By this method a strong desire to study an elementary principle is excited by bringing his labors to a point where he perceives the necessity of it and its direct application to a useful purpose.

" 3. Corporal exercise is not only necessary for the health of students, but for qualifying them for the business of life. When such exercises are chosen by students they are not always judiciously selected.

Such exercises as running, jumping, climbing, scuffling and the like are calculated to detract from that dignity of deportment which becomes a man or science. Therefore a system of exercises is adopted at this school which, while it improves the health, also improves the mind and excludes those vulgarisms which are too often rendered habitual among students. Such exercises as land surveying, general engineering, collecting and preserving specimens in botany, mineralogy and zoology, examining workshops and factories, watching the progress of agricultural operations, making experiments upon nutritious matters proper for vegetables in the experimental garden, etc., are made the duties of students for a stated number of hours on each day."

To further illustrate the methods employed an account will be given of the routine work during the three terms which composed the year. Each term was divided into sub-terms three weeks in duration. Students were admitted at the beginning of any sub-term and their annual course was completed at the end of a year from the time they commenced. The exercises were so arranged that it was a matter of indifference at which sub-term they began. The fall term opened on the third Wednesday in July. The first sub-term was devoted wholly to botany, and each student gave fifteen extemporaneous lectures on this subject before his fellow-students and one or more professors. At the end of the first sub-term the class was distributed into four divisions. The first division was placed in the natural history room for one sub-term, the second in the common laboratory, the third in the natural philosophy room, and the fourth in the assay room.

The equipment of these laboratories, as first established, is interesting : "The natural history room is furnished with sufficient specimens for illustrating mineralogy, botany and zoology, a large furnace, a goniometer, a megascope, a blowpipe, scales, tests, etc, sufficient for investigating subjects in natural history.

" The common laboratory is furnished with a cistern, furnace, and everything necessary for performing chemical experiments, excepting those which teach the analysis of metalloids, metals and animal and vegetable matter.

" The natural philosophy room is furnished with a small observatory, skylights, mechanical powers, hydraulic instruments, optical instruments, mathematical instruments, pneumatical apparatus, etc., sufficient for demonstrating every principle in experimental philosophy.

" The assay room is furnished with skylights, a forge, large bellows, and other conveniences for the analysis of minerals, mineral waters, and animal and vegetable matter."

Each of the four divisions was wholly employed with the subjects assigned to the room occupied by it during one sub-term. Then all the divisions moved on " in a circle ". The first took the place of the

second, the second that of the third; the third that of
the fourth, and the fourth that of the first. At the
beginning of the next sub-term all the divisions
moved on in the circle again as before; and so on,
until each division had devoted a sub-term to each
department.

There was a regular daily routine for the work.
The first bell rang at sunrise and the second twenty
minutes later. Five minutes after this the students
gathered in the reading-room for an examination on
the exercises of the preceding day. At nine o'clock
a lecture was given by a professor to all of the students, and at ten o'clock the daily assistant, called
the officer of the day, gave a lecture before all of them
in the presence of the professor. The place of daily
assistant was filled by the students in rotation. At
the close of the lecture the students criticised his
style, manner and experimental illustrations. Ten
minutes after the close of this exercise two sub-assist-
ants gave lectures in separate rooms, each before
two divisions, in the presence of a professor or assistant. Every one took notes, for use at the meeting
held for purposes of general criticism at the close of
the exercises of the forenoon. At the expiration of
ten minutes from the end of these lectures the four
divisions separated, each going to its respective department, where every student in turn lectured before
the others and a professor or assistant. They then
all met in the reading-room and each criticised all the
lectures he had heard. These exercises closed at

one o'clock. After dinner the divisions went to their respective departments to prepare for the experiments and demonstrations of the next day. After this preparation, which was generally completed by four o'clock, the students met in the reading-room to receive directions for the "afternoon amusements". They were then arranged in divisions and led by professors or assistants to workshops, factories, etc., "for the purpose of applying the principles of mechanical philosophy and chemistry to the various operations of artists," or to the field to collect plants. Five days of each week were occupied as above described. Every other Saturday, and also Friday and Saturday evenings, were devoted to parliamentary exercises. The students represented the different states and formed a parliament for purposes of debate. On the alternate Saturdays not devoted to debate, after the morning examinations were over, they were free for the rest of the day.

The exercises of the Winter term, which was twelve weeks in duration, were conducted on the same plan as that described for the Fall term. Rhetoric, logic, etymology, history, geography and mathematics were taught. The afternoon amusements, adopted according to the state of the weather and without systematic order, were: use of the sextant, compass, goniometer, blowpipe, telescope and other optical instruments, construction and use of ice lenses and prisms, map-drawing and the dissection of animals.

The first six weeks of the Spring term were devoted to a review of the subjects of the Fall term,
and the last nine weeks, or three sub-terms, were
employed in the practical application of the work of
the Fall term.   Instruction was given in the analysis
of selected specimens of minerals, mineral waters,
soils, manures and animal and vegetable matter,
animal and vegetable physiology, origin and nature
of the nutritious substances necessary for the growth
of plants, microscopic examination of the structure
of organized substances, principles of astronomical
calculations, with practical application to eclipses and
matter found in the common almanac ; taking latitude and longitude, lunar observations, etc.   The
afternoon amusements for the last nine weeks were :
collecting and preserving plants, animals and minerals ; land surveying and levelling ; calculating
water pressure in locks, aqueducts, mill flumes, dams,
raceways, penstocks and pumps ; applying the principles of " mechanical philosopy " to the machinery
of steamboats, mills, factories, etc.; application of
mathematics to cask and ship gauging and to other
cases of practical mensuration ; examination of the
progress of agricultural and horticultural operations ;
application of active substances to plants in the experimental garden, such as the strong acids and
alkalies, the various gases, free and combined, and
the effects of the atmospheric gases where all other
active agents are excluded.

Examinations were held at the end of each term ;

GYMNASIUM.   COMPLETED, 1887.

and at the annual examinations in June candidates for degrees gave lectures on the application of the sciences to the common purposes of life. Degrees were conferred annually on the last Wednesday in June.

The system of instruction thus outlined was undoubtedly novel in certain particulars. Its author or authors stoutly maintained that this was the case. Claims for its originality were made in a number of the early circulars./ It is believed that Professor Eaton was responsible for the composition of most of these. We find under the head of "remarks," in a prospectus issued in 1827 : " It will appear from a perusal of this pamphlet that this school is not Fellenbergian nor Lancastrian, but is purely *Rensselaerean.* The unwillingness to admit the *possibility* of an American improvement in the course of education which generally prevails, and the universal homage paid to everything European, has caused much effort to trace the Rensselaerean plan to some supposed shade of it on the other side of the Atlantic. Hitherto these invidious efforts have totally failed." Also : " These principles have now been practically applied for three years, to the full satisfaction of the patron and trustees. The method of teaching by lectures is original ; though Captain Basil Hall, of the British navy, who is now making a tour of the United States, told me that Prof. Pillans of Edinburgh had accidentally fallen upon that method in some degree, though he had received no account of

this school, and that he set a high value upon it."
Again, in a circular issued in 1833 there appears the
paragraph : " It is well known that numerous colleges
(literary and medical), academies, male and female
seminaries, etc., now adopt the experimental method
to a greater or less extent.   Their not acknowledg-
ing the origin of these improvements can never affect
the feelings of the patron.   It is sufficient for his
purpose that the cause of education is improved and
improving by his silent efforts ; without show or loud
pretentions."

The method of instruction pursued by Eaton was
certainly neither that of Lancaster nor of Fellenberg,
though it had points of similarity to both.   His
"officer of the day" performed some of the duties
of the monitor in the Lancastrian system, both
having charge of the classes for a certain period of
the day ; but here the similarity between the two
methods ended.   A short sketch of Fellenberg's
efforts in the cause of education will indicate the difference between his schools and that of Van Rensselaer.   Both men were actuated by the same mo-
tives—the education of those who could not afford
to pay much for the privilege.

Emanuel de Fellenberg was a Swiss nobleman
who, after taking part in the public affairs of his
country during its occupation by the French, determined to devote his life and fortune to the instruction
of the poor.   In 1799 he purchased an estate at
Hofwyl, in the canton of Berne, upon which he estab-

lished his schools for this purpose. His "Agricultural Institution" or "Poor School" was founded in 1808. The fundamental principles in its government were the employment of agriculture for the moral education of the poor and the defrayment of the expense of their education by means of their own labor. About the same time a school of "Theoretical and Practical Agriculture" for all classes was formed. These were very successful, and he soon afterwards commenced the formation of a normal school or seminary for teachers at his own expense. Forty-two teachers of the canton of Berne came together the first year and received a course of instruction in the art of teaching.

In 1827 he established his "Intermediate or Practical Institute", designed for the children of the middle classes of Switzerland. The course of instruction included all the branches which were deemed important in the education of youths not intended for the professions of law, medicine or theology. The pupils belonged to families of men of business, mechanics, professional men and persons in public employment whose means did not allow them to give their children an education of accomplishments. In addition to an ordinary scholastic course the pupils were all employed two hours each day in manual labor on the farm, in a garden plot of their own, in the mechanic's shop and in household offices, such as taking care of rooms, books and tools.* It is there-

* American Journal of Education, Henry Barnard, Vol. III, Hartford, 1857.

fore evident that a marked difference existed between any of Fellenberg's institutions and the R ens-
selaer School.

The practical demonstration of the success of the
system adopted in the experimental school determined the patron and trustees to extend its usefulness
by the establishment of what was called a " preparation branch ", to accommodate those who were dis-
qualified for entrance to the school proper either by
want of education or because they were under seventeen years of age.   It was a preparatory school for
the regular course, and the exercises were of the
same character though more elementary than those
of the latter.   A special laboratory was provided for
this class.   The studies to be pursued and other information relating to it were given in a circular dated
September 14, 1826, which will be quoted in full.

PREPARATION BRANCH RECENTLY ESTABLISHED AT RENSSELAER SCHOOL.

From a respect for the frequent solicitations of many gentlemen in the Southern States, and of some in the Northern, and
from a desire expressed by the patron, to see the results of an
extension of his plan, a *preparation branch* was this day established at this school, to go into operation on the third Wednesday in November.

The following is an outline of the Plan.

1. The original method of instruction which has produced
such unexpected results, called the Rensselaerean method,
will be extended to this branch; to wit, that of exercising the
student, on the forenoon of each day, by causing him to give
an *extemporaneous dissertation* or *lecture* on the subject of his

course, from concise written memoranda; and to spend the afternoon in *scholastic amusements.*

2. The circle of instruction is divided into five parts; and to each part is attached a course of *summer* and *winter* afternoon amusements. The following order will be observed in the fall and winter terms. In the spring term it will be inverted.

First Division. BOTANY and ETYMOLOGY. (The latter branch will extend to so much knowledge of the structure of the Latin, Greek, and French languages, as will enable the student to trace scientific terms to their themes, which are derived from those languages.) AMUSEMENTS. For *summer.* Collecting and preserving minerals, plants and insects. For *winter* none, as this division will not be studied in the winter.

Second Division. GEOGRAPHY and HISTORY. AMUSEMENTS. For *summer.* Selecting specimens for illustrating the physiology of vegetation, and examining them under the common, and the solar, microscopes, and making drawings of their internal structure. For *winter.* Each making a globe of plaster of Paris, and drawing the chief subjects of geography upon it.

Third Division. Elements of PRACTICAL MATHEMATICS and of MORAL PHILOSOPHY. AMUSEMENTS. For *summer.* Land-surveying, taking the latitude, and performing simple hydraulic experiments. For *winter.* Making and using a set of mechanical powers, exercises in percussion with suspended balls, gauging, measuring cordwood and timber.

Fourth Division. LOGIC and RHETORIC. AMUSEMENTS. For *summer.* Experimenting upon the most common gases, as oxygen (obtained from vegetables by the action of light), nitrogen, hydrogen, carbonic acid (with its combination in soda-water), testing their specific gravities, etc., and experimenting upon aqueous exhalations—all to be performed with apparatus made with their own hands. For *winter.* Making and using galvanic batteries and piles, electrometers and magnets; and disengaging combined caloric by compression and affinity.

Fifth Division. Elementary principles of GOVERNMENT and

LAW, and PARLIAMENTARY RULES.   AMUSEMENTS.   For *spring*
and *fall.*   Constructing dials, fixing meridians, constructing
and using air-thermometers and hygrometers, taking specific
gravities, using the blow-pipe and constructing the three elementary musical chords to illustrate the science of tones.
For *winter.*   Making camera-obscura boxes; producing focal
images by a pair of common burning glasses and ice lenses,
and illustrating the microscope and telescope by the same ;
illustrating the laws of refraction and reflection by cheap mirrors and vessels of water, and separating the colored rays by
ice cut into triangular prisms.

Candidates are admitted to the preparation branch, who are
deemed of sufficient discretion for going through the course,
provided they have been successfully taught in reading, writing, common arithmetic and English grammar.   The Faculty
of Rensselaer School are to judge upon their qualifications;
but the Trustees have, in the second article of the by-laws of
this branch, expressed an opinion, that "the age of thirteen or
fourteen years and upwards, is best adapted to this course."

EXPENSES.   *Tuition* $1.50 for every three weeks, which
constitutes a step in the circle.   Students may enter either
step in the circle at the commencement of every three weeks,
reckoning from the beginning of each term.   The terms
or sessions of this branch, correspond with the other terms
of the School.   *Board,* in commons with the other students,
never to exceed $1.50 per week.   Rooms will be furnished
at or near the school, to be under the inspection and control of the faculty, at a small expense.   No charge is made for
the use of public rooms, library, chemical and philosophical
apparatus, tools of the workshop, or the cabinet.   And each
student will attend the daily lectures of the Professors, free of
charges.   A student of strict prudence, may pay all his expenses for the 42 weeks in each year, at this branch, with
$120, as follows : Tuition $21: board $63: fuel and lights
$10: washing and lodging $10: text books $6: amusement
apparatus, $10.

As this circular may fall into the hands of some, who have
not read the new code of by-laws, passed April 3d, 1826, and

the legislative act of incorporation, passed March 21st, 1826, it may be advisable to state as follows:

The Rensselaer School was founded by the Honorable Stephen Van Rensselaer, solely for the purpose of affording an opportunity to the farmer, the mechanic, the clergyman, the lawyer, the physician, the merchant, and in short, to the man of business or of leisure, of any calling whatever, to become *practically scientific.* Though the branches which are not taught here, are held in high estimation, it is believed that a school attempting every thing, makes proficients in nothing. The Rensselaer School, therefore, is limited to an EXPERI-MENTAL COURSE in the NATURAL SCIENCES. The studies of the preparation branch are extended no farther than is neces-sary, as auxilaries to the experimental course.

The FALL TERM commences on the third Wednesday in July, and continues 15 weeks.

The WINTER TERM commences on the third Wednesday in November, and continues 12 weeks.

The SPRING TERM commences on the first Wednesday in March, and continues until the last Wednesday in June; which is the day of the annual commencement.

EXPENSES. All the same as in the preparation branch, with the addition of double the charge for tuition in the fall and spring terms, on account of the great additional labor required for teaching the student to perform with his own hands about sixteen hundred experiments in chemistry and natural philos-ophy. But students who have gone through a course in the preparation branch with success, will not be required to at-tend the winter term. This will reduce the necessary ex-penses to about $95 for the whole experimental course.

Many unsuccessful attempts have been made to render science amusing to the youthful mind. They have generally proved very unprofitable, by diverting the attention of the student from literary pursuits, and by creating an attachment to useless, and often demoralizing sports. By the plan adopted at this school, the objections to scholastic amusements are ef-fectually obviated; and it will appear by this circular, that those have been selected, which will give due exercise to both body and mind. The muscular powers of the body will be called into action, and their forces will be directed by mental ingenuity, until the student becomes familiar with the most important scientific manipulations, and particularly with those which will be most useful in the common concerns of life.

The Rensselaerean scheme for communicating scientific knowledge had never been attempted on either continent, until it was instituted at this school, two years ago. Many indeed mistook it, at first, for Fellenberg's method; but its great superiority has now been satisfactorily tested by its effects. As the *experimental school*, as well as the *preparation branch*, were founded solely for the public benefit by its disinterested patron, it is the particular desire of the trustees, that its excellences should be understood and imitated at other schools, as set forth in a former circular. Like other useful inventions, much expense was required for making the first experiment. Fortunately for science, the trial has been fairly made at the expense of many thousands, advanced by a single individual. Now it may be followed, in its chief advantages, by every school district ; while the parent school at Troy will prepare competent teachers.

By order of the Trustees.

SAMUEL BLATCHFORD, *President.*

RENSSELAER SCHOOL, TROY, (N. Y.) Sept. 14, 1826.

CHAPTER V.

THE NAME CHANGED TO RENSSELAER INSTITUTE.
REMOVAL TO THE VAN DER HEYDEN MANSION.

It has been seen that from its beginning an essential part of the educational system of the school consisted in an examination of workshops and factories in the neighborhood of Troy and in botanical and geological excursions in its immediate vicinity. It was determined to extend such excursions to more distant points in order to afford better facilities for the practical study of mineralogy and geology. At a meeting of the board of trustees held February 12, 1827, a by-law was passed requiring each student to make "a tour of about three weeks along the transition and secondary district of the Erie Canal immediately after commencement and across the primitive district in an eastern direction immediately after the close of the fall term." In a circular of six pages, written by Amos Eaton, entitled, "Rensselaer School Flotilla for the Summer of 1830", the programme of a proposed travelling tour for that year is given in detail. It was to begin on the twenty-third of June and to last ten weeks. Students taking it were to meet at the dock at the lower end of Cortlandt Street in New York City and to proceed by steamboat to Albany, whence a flotilla of canal-boats was to take them through the Erie Canal to Lake Erie. They were to return by the same route. Daily lectures were to be given in the morning, and in the afternoon botanical and geological excursions were to be made. The boats were to move slowly so that specimens could be obtained at any point along the route. There is a list of twenty-nine places to be visited, Trenton Falls, Niagara Falls and Lockport being included. This trip was not obligatory, and in succeeding publications three excursions which might be substituted for it are enumerated : one to the Connecticut River, one to the Helderberg and the third to Carbondale, Pa., and Amboy, N. J.

At this time the total cost of attendance for one year, including excursions, was said to be $230, though it was observed that a young gentleman of tolerable economy could reduce this to $170.

At the trustee meeting to which reference has just been made there was also added to the curriculum the requirement that students should speak extemporaneously once a week during the winter term and twice a month during the other terms.

At the same time the first " prudential committee ", consisting of the president and two trustees, was appointed. Succeeding boards have retained this committee, which has the power to perform, between the regular meetings of the board, such duties as cannot properly be delayed.

To further increase the usefulness of the institution

ALUMNI BUILDING.   COMPLETED, 1893.

the faculty were authorized, May 24, 1827, to establish district branches in any part of the state when
application was made and assurance given by responsible persons that suitable rooms and sufficient
apparatus would be supplied. The object was to accommodate those who wished to be educated and
yet were unable to leave home for the whole or even
a part of the year. It was provided that the branch
students were to be taught that part of the annual
course which did not require expensive apparatus;
" for more than three fourths of an experimental course
of scientific instruction may be taught with apparatus
worth but one hundred dollars; whereas the remaining fourth requires apparatus worth three or four
thousand dollars." Should they desire, they might
then come to the school, and after devoting nine
weeks to that part of the course requiring expensive
apparatus, they would be received as candidates for
the Rensselaer degree on an equal footing with those
who had spent the whole year at Troy.

Complete directions for introducing experimental
science in academies and common schools were also
given at this time. Beside information in relation to
the regular work to be pursued, advice was given
regarding the "amusements". Under this head oc-
curs the clause : " A level sufficiently accurate may be
made by any one, with the cost of a spirit-level tube
of but a few shillings' value. Such students may
then be taught the general outlines of civil engineering, land surveying, etc., in lieu of mischievous tricks,

degrading contortions called gymnastics and profane language." The circular from which this quotation is taken was dated September 19, 1828. There is added to it a note in which we are informed that forty mechanics, members of the Mechanics' Institute of Troy, placed themselves under the direction of the Rensselaer School during the winter of 1827, and that most of them became tolerably proficient in experimental chemistry as applied to the arts and manufactures. They were not regular members of the school but paid one of the professors to teach them.

All these efforts show the active interest displayed by the founder and the officers of the school in the extension of the experimental system and the diffusion of scientific knowledge. (To extend still further the benefits of the institution Mr. Van Rensselaer, while in the House of Representatives, wrote from Washington the following letter to the president of the Institute. It was dated December 31, 1827 :

" *Dear Sir:* I take the liberty of suggesting to you and the trustees the propriety of offering the school (over which you preside with so much dignity and usefulness) to the Legislature, to educate teachers, as proposed by Gov. Clinton in his message at a former session of the Legislature—perhaps an amendment to the charter, extending the power of the trustees to change the location of the School, if they deem it necessary."

Nothing having come from this suggestion, he

caused, in 1828, an invitation to be given to each county of the state to furnish a student, selected by the clerk of the county, for gratuitous instruction at Troy. This invitation was accepted by nearly all the counties. The students thus instructed were required to teach the experimental and demonstrative method in their own counties for a period of one year.

The authorities of the school seem also to have had, for those days, advanced ideas in regard to the education of women, for we find, as an addendum to a circular dated October 29, 1828, the following " *Notice by A. Eaton, in his private capacity.* ' At the urgent solicitations of several judicious friends, a lady, well qualified for the duty, will take charge of two experimental courses in chemistry and natural philosophy, in each year, for ladies : similar to the courses proposed for gentlemen in the annexed circular. They will be nine-week courses, at the same times and for the same charges. But no extemporaneous lectures will be required, excepting of those ladies who wish to prepare for giving instruction '."

And in the minutes of the board there is a copy of a letter from Professor Eaton to the examiners, dated February 11, 1835, in which he requests them to give an informal examination to eight young ladies, who had been instructed for one quarter in practical mathematics, "so far as to be enabled to draw a fair comparison between the study of speculative geometry and algebra as generally practised in female seminaries and this mode of applying mathematics to the essential calculations of geography, astronomy, meteorology, necessary admeasurements, etc." The examiners complied with his request and were highly gratified at the progress made by the class.)

It may be explained that all examinations, in the early period of the school's history, were made by boards composed of from three to six qualified persons appointed by the trustees. None of the members of these boards was connected with the school.

(Professor Eaton's pronounced opinions upon the educational methods generally pursued in schools for young men have been illustrated in preceding pages. These extended to the education of women as well, and the manner in which he expressed them was quite as forcible in the one case as in the other. He remarks, at the end of a printed synopsis of the mathematical course for the year 1834–5 : " The waste of time in many female schools, by the fashionable mummery of algebra, half learned and never applied, has caused many to ascribe the failure in mathematics to the perversion of female genius, when it is drawn from elegant literature, music, painting, etc., to the severe sciences. The true cause is to be found in parsimony, which excludes competent teachers, badly selected subjects and wretchedly compiled text-books.) Our country is inundated with wild schemes of learning ; while the speculating book-sellers are sending their harpie-like pedlars to rob our youth of the last fragments of common sense."

Although by the year 1829, after a trial of four years, it had been conclusively proved that the experimental and demonstrative method, as they called it, was successful as a system of instruction, the institution had not been self-supporting. Its founder paid each year more than one half of its expenses. This was becoming burdensome to him, and he signified to the trustees his desire to discontinue it, and especially his intention of discontinuing the gratuitous education of county students after October, 1829. He did not in fact cease to contribute to the support of the school, but in consequence of this declaration it was " farmed out " in November, 1829, to Amos Eaton for a period of one year. He was constituted the " Agent " of the trustees to transact all the pecuniary business of the institution, which, however, was to remain under the control of the board. He relinquished all claim for compensation, and in consequence was authorized to receive and expend all moneys at his discretion and to retain all profits for his own benefit. An inventory of the property was made and he was permitted to use it for purposes of instruction. This arrangement was continued for one year only, as he terminated it in September, 1830, although he still acted as agent and retained his position as Senior Professor.

In spite of pecuniary embarrassments, improvements were continually being made both in the instruction and the equipment of the laboratories. The prospectus for the eighth annual course shows that in

1831–2 the year had been divided into seventeen sub-terms of three weeks each, of which, however, three, called "reading terms", might be used either to visit friends or for a course of reading in the library. The fifteenth and sixteenth sub-terms were occupied in the travelling tours to which reference has been made.

During the morning exercises of the year each student had to give one hundred and eighty extemporaneous lectures, upon which he was closely criticised. These lectures were illustrated by about twelve hundred experiments performed by himself, and by "suits" of minerals, plants and animals.

At this time the equipment included a reading-room, a natural history room, a philosophy room and three laboratories. Considerable additions had been made to the apparatus as described in the circulars of 1826. The philosophy room now contained an air-pump, a force-pump, barometer, thermometers, pluviometer, solar microscope, megascope, standing microscope, magic lantern, telescope, lenses, convex and concave mirrors, prisms, electrical-machine, galvanic battery, electromagnetic instrument, magnets, sextant, theodolite, compass and chain, mechanical powers, hydrostatic bellows, hydrostatic and hydraulic cylinders and tubes, hydrometers and glass pumps.

The laboratories were furnished with the necessary forges, furnaces, bellows, lead-pots, Argand lamps, common lamps, iron retorts or gun-barrels for gases, anvils, anvil hammers, cisterns, pipes for conducting

VAN DER HEYDEN MANSION. OCCUPIED, 1834-41.

gases from the barrels, gas-pistol, iron stand, iron mortar and mercurial bath.

In the meantime the Rev. Samuel Blatchford, after earnest and successful labor in behalf of the school, died March 27, 1828, and was succeeded by the Rev. John Chester, a clergyman of Albany, who was appointed June 25, 1828. His term was, however, a short one, as he was compelled, on account of ill health, to resign in about six months. He was succeeded by the Rev. Eliphalet Nott, appointed September 2, 1829, who was at the same time president of Union College.

During the first seven years of its existence the school had been situated at the corner of Middleburgh and River streets, in the building formerly occupied by the Farmers' Bank, and known, at the time of its establishment, as the Old Bank Place. Partly because it had not yet become self-supporting and partly because it was, in some respects, not conveniently situated, it was determined to obtain authority from the legislature to change its location if satisfactory arrangements could be made. An act was consequently passed April 26, 1832, which gave the trustees power, after October 23, 1832, if the patron consented, to remove to the site of the Greenbush and Schodack Academy, in the town of Greenbush, in Rensselaer county, and to unite with this academy if its trustees consented. In this case the united institution was to be called the Rensselaer Institute. If, however, the patron or the trustees of the academy objected, the trustees of Rensselaer School were given authority to remove the institution, after the consent of Stephen Van Rensselaer had been given, to any part of Rensselaer county and to continue as an experimental and classical school under the name of the Rensselaer Institute.

The inquiries and negotiations made, in relation to the removal to Greenbush, were not satisfactory, as may be seen from the following letter written by the patron to the Rev. Dr. Nott and read at a meeting of the board of trustees held November 18, 1833 :

" ALBANY, November 18, 1833.

" *To the President and Trustees of the Rensselaer School:*

" *Gentlemen:* Sufficient provision for the support of said school not being offered to its location at Greenbush, according to the first section of the amendment of April 26, 1832, I feel bound in duty to object to its removal to Greenbush. But under present circumstances I cheerfully consent to a removal to the Van der Heyden mansion, or to any other suitable building near the central part of said city of Troy.

Respectfully your humble servant,

"S. V. RENSSELAER."

Among the by-laws passed at this meeting was one by which the name of the school was changed to the " Rensselaer Institute," which was to include an " experimental and classical department ". At the same time the scholastic year was divided into two terms instead of three, the winter term, sixteen weeks in duration, to commence on the third Wednesday in November; and the summer term, of twenty-four weeks, to begin on the last Wednesday in April. Each term was divided into sub-terms of four weeks each. It was also resolved to remove to the Van der Heyden mansion on or before April, 1834. This building was selected on account of its size and convenience of access. It was situated on the southwest corner of Eighth and Grand Division streets, and the removal took place in April, 1834.

During the occupation of the Old Bank Place the number of students at any one time had never exceeded and was generally less than twenty-five. The number of teachers was regulated by the number of students, one being assigned to each section of five or six. The triennial catalogue for 1832–3–4 gives a list of twenty-five instructors who had already been connected with the school. The small number of students was partly due to the standard required for entrance to the regular course; at one time twelve of the twenty-five present were graduates or members of colleges. In the notices for the ninth annual course, 1832–3, during the time that the change of location was being considered, it is remarked: " None are received but those whose minds are disciplined to habits of study. Hence it is that the patron has already advanced over twenty-two thousand dollars in support of the school for eight years. To improve the plan of education is his object; not to establish a school at any particular location. Therefore patronage is not asked. These terms are printed, not for the benefit of the school, but for the benefit of those who wish to profit by the improvements made by trials which cost the patron many thousands."

The first clause of the preceding quotation could hardly have referred to the junior members of the school, in the Preparation Branch; as Rule 8 of the by-laws of 1835 reads : " In case of any disobedience of any juniors to orders of teachers, after being particularly called to obey, it shall be the duty of said professor to lay hands on such disobedient student and remove him from the premises, or confine him (in such a manner as to cause no personal injury) for a time not exceeding two hours. But no beating or flagellation shall in any case be permitted at the Institute."

CHAPTER VI.

ESTABLISHMENT OF THE DEPARTMENT OF CIVIL ENGINEERING.

THE preceding pages show that the original intention of the founder was to establish a school for the diffusion of scientific knowledge, and that his object more particularly was to disseminate among farmers, mechanics and the poorer classes generally information in relation to the application of scientific principles to their various occupations which would enable them to improve their material condition. At the same time the management of the institution was of too broadminded a character to permit its benefits to be confined to any particular branch of practical science, and, although many of those who had up to this time been graduated afterward became eminent in various departments of pure and applied science, the renown of the school is principally due to the work of its alumni in the field of engineering—a course in which was about to be added to the curriculum.

No school of civil as distinguished from military engineering had yet been established in any English speaking country, although on the continent of Europe a number of technical institutions had been founded, most of which were maintained partly or wholly at the expense of the state. The École des Ponts et Chaussées was established in France as early as 1747, though it did not become of importance as a school for engineers until a much later period, and the Königliche Sächsische Bergakademie (Freiberg) was founded in 1765. Among other continental technical schools of early date which afterwards became well known may be mentioned the École Polytechnique (Paris, 1794), a school of general science, having for its principal object the preparation of students for several special government technical institutions, including the School of Bridges and Roads above mentioned; the Polytechnisches Institut (Vienna, 1815), intended for the education of engineers, architects and manufacturers; and the Königliches Gewerbe Institut (Berlin, 1821), which at the time of its foundation and for twenty-five years thereafter was, as its name indicates, a trade rather than an engineering school. The Technische Böhmische Ständische Lehranstalt (Prague) came into existence in 1806. Beside these, which depended largely upon government aid, a private institution, the École Centrale des Arts et Manufactures (Paris, 1829), attained prominence as a school of engineering immediately upon its establishment. Before 1835 a few other technical schools of less importance, containing trade-school features, had been founded in the German states.

The continental schools of science antedated those of Great Britain. Among the English schools in

USING THE SOLAR TRANSIT. 1894.

which scientific instruction was early given may be mentioned University College, London, which was opened in 1828 under the name of the University of London and King's College, London, established by royal charter in 1829. In the University of London engineering subjects were first taught in 1840 ; and in the same year a chair of civil engineering and mechanics was established by Queen Victoria in the University of Glasgow. The School of Engineering in Dublin University (Trinity College) was founded in 1842. The other well-known British schools of science were established at still later dates. Among them are Owens College, Manchester (1851) ; the Department of Engineering in the University of Edinburgh (1868) ; the Royal Indian Engineering College, London (1871) and Mason College, Birmingham (1875).

In this country the Military Academy at West Point, which was established in 1802, though it was a school in name only until its reorganization after the war of 1812, was the only institution giving an education to which the word engineering could be applied, and it, of course, was a military school.

In fact, at the time of the foundation of Rensselaer School it could scarcely be said that there were any engineers other than military engineers. The term civil engineer had only recently come into existence. There were no schools of civil engineering here because, although there had been inventors and constructors of genius before that date, civil engineering had hardly yet been recognized as a profession. A consideration of the condition of the country and of the state of scientific knowledge as applied to the constructive arts towards the beginning of the century shows why this was the case. In comparison with the European states, in which the early schools of science above mentioned had been established, the country was new and sparsely settled. In the year 1800 the total population of the United States was only 5,300,000. In the same year the state of New York contained 589,000 and New York City only 60,000 inhabitants. In 1830 the country had 12,866,000 inhabitants, while New York state had 1,919,000 and New York City 203,000. Troy was a village of 1800 people at the former period, and in 1830 this number had increased to 11,500. Methods of communication were primitive and travelling was expensive.

No canal of considerable length (and these were the first engineering works of great magnitude to be built) was begun until after the conclusion of the second war with England, that of the Schuylkill Coal and Navigation Company, 108 miles in length, being commenced in 1816 and finished in 1825. Others in Pennsylvania were commenced about the same time, and both the Erie and Champlain canals were begun in 1817. By the end of the first quarter of the century about 1400 miles of these waterways had been built; but no steam railroads existed, locomotives not becoming practically successful until about 1830. The first ones used weighed only three or four tons, although in the years 1836–7 Baldwin of Philadelphia built eighty weighing from nine to twelve tons each.

Steam navigation was in a more forward state : the Clermont, a steamer one hundred and thirty-three feet in length, built by Fulton and Livingston in 1807, having made the trip up the Hudson River from New York to Albany in thirty-two hours. A steam ferry-boat ran between Jersey City and New York in 1812, and in 1815 there were steamboats running between New York and Providence. In the year 1830 there were eighty-six steamers on the Hudson River and Long Island Sound. The first steamship to cross the Atlantic was the Savannah, of 350 tons, built at Corlears Hook, N. Y. The engines, however, were used only eighteen out of the twenty-five days required for the passage from Savannah to Liverpool, and sails had to be depended upon for the remainder of the trip. It was not until 1838 that the transatlantic voyage was made wholly by steam. In this year the Sirius, of 700 tons, crossed from Cork to New York in nineteen days, and the Great Western, of 1340 tons, made the passage from Bristol to New York in fifteen days.

In the early days of the country the small amount of power required for manufacturing purposes was obtained principally from wind and water wheels. Of the latter, undershot, overshot and breast wheels were employed; and Francis says that until 1844 high-breast wheels were considered the most perfect water-wheels that could be used. Although Four-neyron had erected his first turbine, in France, in 1827, and Elwood Morris of Pennsylvania had shortly afterwards built and put two of them in operation in this country, other wheels of this type were not used here until about the middle of the century. Boyden designed his turbine in 1844; and the Manufacturing Companies at Lowell, which had begun to improve the water-power of the Merrimac in 1822, purchased the right to use it in 1849.

The practical application, in Great Britain, of the steam-engine to pumping water from mines led to the introduction of the first one of any size ever used in America. All of its principal parts were imported from England and a mechanic was sent over to erect and run it. It was put together in 1763 at the Schuyler copper-mine on the Passaic River, a few miles above Newark, N. J. Frederick Graff says[*] that in 1803 there were in use in the United States five steam-engines beside the one referred to above; two at the Philadelphia water-works, one just about being started at the Manhattan water-works in New York, one in Boston, one in Roosevelt's saw-mill in New York, and quite a small one used by Oliver Evans to grind plaster of paris, in Philadelphia. The first steam-engine built in America is said to have been constructed in 1772 by Christopher Colles for a

[*] Notice of the Earliest Steam-engines used in the United States, by Frederick Graff, in Journal of the Franklin Institute, 1853.

distillery in Philadelphia, but it was very defective. Those of the Philadelphia water-works were built in 1800 at the Soho Works of Roosevelt, near Newark, N. J. From this time onward the application of steam as a source of power for manufacturing purposes increased with the demands of the times. Improvements—dictated by experience, for little was known of the theory—were continually made, and by the middle of the century the various types had assumed practically the proportions used at the present time.

One of the first tunnels built in the United States was on the Allegheny Portage Railroad in Pennsylvania. It was built in 1831 and was 900 feet long. The Black Rock tunnel on the Reading Railroad was built in 1836. It was 1932 feet long. In 1820 one of the first cast-iron water-mains in the country was laid for the Philadelphia water-works.

Bridges of wood and stone had of course been built almost from the time of settlement of the country. Some of the former were of long span and reflected the greatest credit upon the genius of their constructors, who, however, had only empiric methods of proportioning the parts. Palmer, Burr and Wernwag were the most noted builders at the beginning of the century. The Piscatauqua bridge, built by Palmer, near Portsmouth, N. H., included an arch span 244 feet in length; and his Schuylkill River bridge had two arch spans 150 feet and one 195 feet long. Between 1804 and 1808 Burr built his Waterford, Trenton and Schenectady bridges, with spans ranging from 150 to 203 feet, and, from 1812 to 1816, the Harrisburgh bridge, with twelve spans of about 210 feet each. Wernwag built his "Colossus" over the Schuylkill at Philadelphia in 1812. The span was 340 feet. Town patented his lattice truss in 1820, and Howe's patent was not taken out until 1840. The era of iron bridges did not begin until 1840. Finley had built a number of small suspension, bridges of chain cables between 1796 and 1810; and in 1810 Templeman replaced the 160-foot span of Palmer's Essex-Merrimac bridge by one of chain cables. Paine's memoir on cast-iron bridges was printed in 1803, and Canfield took out the first patent for an iron truss bridge in 1833; but the first iron truss bridge built in this country is believed to be the one erected in 1840 by Trumbull over the Erie Canal at Frankford.* In the same year Whipple built his first iron bridge.)

The few historical facts above given serve to indicate the condition of engineering science at the period of the school's history which we are now considering. Although many of the fundamental principles of applied mechanics were known as well then as now, the development of the science, particularly in its application to structures and machines for the production of useful work, had taken place largely upon empiric lines. Most of the eminent men to

---

*American Railroad Bridges, by Theodore Cooper, in Transactions of the American Society of Civil Engineers, July, 1889.

IN THE CHEMICAL LABORATORY.   1894.

whom this development had been due were self-taught, were mechanics whose results had been obtained by successive experiments and with little knowledge of the resistance of materials or of the principles of the design of engineering constructions as practised to-day. And if with these conditions there is taken into consideration the comparative smallness of the population and its extended geographical distribution, the wise forethought and liberality of mind displayed by the authorities of the school in establishing at such an early date a department of civil engineering will be thoroughly appreciated.

In the pamphlet, published in 1826, giving the constitution and laws of the school, instruction in land surveying was included among the duties of the Senior Professor, and in the catalogue of officers published in 1828 he was required to lecture on land surveying and civil engineering. This is the first appearance of the term "civil engineering" in any of the circulars, and no well-defined course in the subject was formulated for several years. In the "Notices for the Eighth Annual Course", (1831-2), to which reference has before been made, the first sub-term, beginning November 16, was devoted to "Practical Mathematics, including mensuration applied to land surveying, timber and cord-wood measure, excavations, docks, etc.", and the second sub-term, from December 7 to December 28, to "Trigonometry, Navigation and the elements of Civil Engineering". The fifteenth and sixteenth sub-terms, from Septem-

ber 12 to October 24, were occupied in the "application of Engineering and Natural History to the
occurrences of four travelling tours—to Connecticut
River, to the Helderberg, to Carbondale coal beds
and to New Jersey". These quotations include all
references to the subject; and in the "Notices" for
the ninth annual course civil engineering is not speci-
fically mentioned, though this was an octavo circular
containing only three printed pages.

In 1833 the curriculum in the experimental department contained "Practical Mathematics, including
Surveying, Engineering, Navigation, Latitude and
Longitude, etc., from the 3rd Wednesday in November, 12 weeks." In the original minutes of the board
of trustees we find a record of the examinations of
fourteen students in surveying and in engineering.
These were held February 11 and 12, 1834.

Up to this time the degree of Bachelor of Arts,
A.B. (r.s.), was the only one conferred by the institution, and although the course in engineering had been
gradually developing it had not yet been differen-
tiated from that in general science. Preparatory to
the separation of these two branches the legislature
was petitioned to amend the charter of the school.
This was done by an act dated May 9, 1835. The
second section of this law reads as follows: "The
said board of trustees shall have the power to establish a department of mathematical arts, for the
purpose of giving instruction in engineering and
technology, as a branch of said institute; and to re-

ceive and apply donations for procuring instruments and other facilities suitable for giving such instruction in a practical manner, and to authorize the president of said institute to confer certificates on students in said department in testimony of their respective qualifications for practical operations in the mathematical arts."

At a meeting of the board of trustees held May 22, 1835, their number was increased, in accordance with a provision of the above-mentioned act, by the addition of the Mayor, Recorder and Alderman of the Fourth Ward of the city of Troy; and it was resolved that "A department of Mathematical Arts is hereby established as a branch of the Institute for the purpose of giving instruction in Engineering and Technology". At the same meeting it was decided that the degree of Bachelor of Natural Science, B.N.S., should thereafter be conferred instead of Bachelor of Arts, and that graduates in the department of Mathematical Arts should receive the degree of Civil Engineer. Also that "no one shall receive the last-mentioned degree until he shall have been regularly disciplined at this school at least two quarters, after being well taught in elementary mathematics here, or elsewhere".

The first class in civil engineering was graduated in 1835. The first four candidates for the degree were recommended in the following letter from the examiners, dated October 14, 1835:

"To the Revd. E. Nott, D.D., President:

"We. have examined Edward Suffern, William Clement, Jacob Eddy and Amos Westcott as candidates for the degree of Civil Engineer. We find them acquainted with the theory of practice. But as this is the first class proposed to be graduated, their own honor and the honor of this institution demand great caution in conferring degrees. We therefore recommend as follows : that they receive the degrees, but that the diplomas be left with the Secretary until the President shall receive satisfactory certificates that they have reviewed their Text Books (outlines Gregory), that they can read algebraic equations. and have a general knowledge of Perspective generally.

"A. R. JUDAH, *Chairman.*
"P. H. GREEN, } *Examiners."*
"HARVEY WARNER, }

By this time a complete curriculum in civil engineering had been established. It was printed in a circular which will be given in full, as it is believed to be the first prospectus of a school of civil engineering ever printed in English. It is well worth perusal, not only because the curriculum outlined contains much information regarding the most advanced scientific instruction given in this country at that period, but because the concluding paragraphs throw a curious light upon the expenses of students and the general requirements necessary for graduation.

ASSAYING.   1894.

NOTICES OF RENSSELAER INSTITUTE.

TROY, N. Y., October 14, 1835.

[Being the answer to letters of inquiry.]

HON. STEPHEN VAN RENSSELAER, Patron, with the right to appoint the Annual Board of Examiners.

ACTING FACULTY.

Rev. E. NOTT, D.D., President—also President of Union College.

Judge DAVID BUEL, Jr., Vice President.

AMOS EATON, Senior Professor, and Professor of Civil Engineering; also holding the Agency and Supervision of the Institute.

EBENEZER EMMONS, Junior Professor.

JAMES HALL, Professor of Chemistry and Physiology.

Assistants—Edward Suffern and D. S. Smalley.

Instruction, wholly practical, illustrated by Experiments and Specimens, is given 40 weeks in each year. Five days in each week the forenoon exercises are from 8 A.M. to 1 P.M.

WINTER SESSION commences the third Wednesday in November, and continues 16 weeks. During the first 12 weeks, each fornoon is devoted to practical Mathematics, Arithmetical and Geometrical. This is a most important course for men of business, young and old. During the last 4 weeks of the Winter Term, extemporaneous Speaking on the subjects of Logic, Rhetoric, Geology, Geography and History, is the forenoon exercise. Throughout the whole session the afternoon exercises are Composition, and, in fair weather, exercises in various Mathematical Arts. A course of Lectures on National and Municipal Law, is given by the Senior Professor.

SUMMER SESSION commences on the last Wednesday in April, and continues 24 weeks: ending with Commencement.

Students of the Natural Science Department are instructed as follows :

Three weeks, wholly practical Botany, with specimens.

Four weeks, Zoology, including organic remains ; and Physiology, including the elements of Organic Chemistry.

Three and a half weeks, Geology and Mineralogy, with specimens.

Three weeks, traveling between Connecticut River and Schoharie Kill, for making collections to be preserved by each student, and exhibited at examinations ; also for improving in the knowledge of Natural History and Mathematical Arts.

Ten weeks, Chemistry and Natural Philosophy.

Half a week, preparing for examination and Commencement.

The afternoons of all fair days are devoted to Surveying, Engineering, and various Mathematical Arts—also to Mineralizing, Botanizing, and to collecting and preserving subjects in Zoology.

Students of the Engineer Corps are instructed as follows :

Eight weeks, in learning the use of Instruments; as Compass, Chain, Scale, Protractor, Dividers, Level, Quadrant, Sextant, Barometer, Hydrometer, Hygrometer, Pluviometer, Thermometer, Telescope, Microscope, etc., with their applications to Surveying, Protracting, Leveling, calculating Excavations and Embankments, taking Heights and Distances, Specific Gravity and Weight of Liquids, Degrees of Moisture, Storms, Temperature, Latitude and Longitude by lunar observations and eclipses.

Eight weeks, Mechanical Powers, Circles, Conic Sections, construction of Bridges, Arches, Piers, Rail-Roads, Canals, running Circles for Rail-Ways, correcting the errors of long Levels, caused by refraction and the Earth's convexity, calculating the height of the Atmosphere by twilight, and its whole weight on any given portion of the Earth, its pressure on Hills and in Valleys as affecting the height for fixing the lower valve of a Pump ; in calculating the Moon's distance by its horizontal parallax, and the distances of Planets by proportionals of cubes of times to squares of distances.

Four weeks, in calculating the quantity of Water per second, etc., supplied by streams as feeders for Canals, or for turning Machinery ; in calculating the velocity and quantity effused per second, etc., from flumes and various vessels, under various heads; the result of various accelerating and retarding forces of water flowing in open race-ways and pipes of waterworks, and in numerous miscellaneous calculations respecting Hydrostatics and Hydrodynamics.

Four weeks, study the effect of Steam and inspect its various applications—Wind, as applied to Machinery ; also Electro-Magnetism—inspect the principal Mills, Factories, and other Machinery or works which come within the province of Mathematical Arts ; also, study as much Geology as may be required for judging of Rocks and Earth concerned in construction.

Fees for instruction, including all Lectures, Experiments, etc.; also for use of Instruments, Apparatus, Library and Specimens, $4 for each sub-term of four weeks. No student received for less than a sub-term. No extra charge excepting $8 for the course of Experimental Chemistry, where each student gives a course of experiments with his own hands.

Students furnish their own fuel, light, and text-books. Each boards where he pleases ; but the Professors will aid strangers in the selection of boarding houses. A small number of strangers are boarded at the School at $2 per week ; they furnishing their own bedding, washing, etc.

The Rensselaer degree of Bachelor of Natural Science is conferred on all qualified persons of 17 years or upwards. The Rensselaer degree of Civil Engineer is conferred on candidates of 17 years and upwards, who are well qualified in that department. This power was given to the President, by an amendment to the Charter, passed last session of the Legislature. Candidates are admitted to the Institute who have a good knowledge of Arithmetic, and can understand good authors readily, and can compose with considerable facility.

After a trial of two seasons, it is found to be inexpedient to enter young lads in the regular divisions, before they have sufficient pride of character to govern their conduct when preparing for their exercises in the absence of a teacher ; arrangements will therefore be made for having a teacher always present with them, when they are not in the immediate charge of a Professor or Assistant.

Students in any one department have the right to attend one Experimental Lecture each day in the other departments, free of expense.

One year is sufficient for obtaining the Rensselaer degree of Bachelor of Natural Science, or of Civil Engineer, for a candidate who is well prepared to enter. Graduates of Colleges may succeed by close application during the 24 weeks in the Summer term.

Candidates may commence the course at the beginning of any sub-term; but the third Wednesday of November is to be preferred, unless the candidate is a graduate of a regular College, or otherwise well instructed in general Mathematics and Literature. In such cases the last Wednesday in April is the most suitable time of entering. ·His theoretical views may then be reduced to practice during the Summer course.

The degree of Master of Arts is conferred after two years of practical application.

Gentlemen wishing to learn the outline of the terms of the Rensselaer Institute, are requested to pay postage on their letters; and they will receive this printed notice. If this appears to be a " *narrow notice*," I will state that I paid $54.28 in one year in postage for letters on others' business: some for our school course, more for advice about mines, minerals, and visionary projects.

Amos Eaton, Agent.

RENSSELAER INSTITUTE, TROY, Oct. 14, 1835.

A better understanding of the scope of the instruction given may be obtained from an examination paper covering the work of the winter term in the department of Mathematical Arts. This was submitted to fifteen students ; and the results of the examination are given in a report of three examiners, dated February 23, 1836. There were fifty-three questions :

List of Subjects for Examination.

1. Extract the square root.   Illustrate by diagram.

2. Find by the square root the length of a ladder placed against a wall 37 feet high, its bottom being 9 feet from the wall.

3. Demonstrate this application of the square root by trigonometry.

4. Find the distance across a river without instruments, by calculating a base frustrum of an isosceles triangle, pointing the apex to an object on the opposite shore.

5. Explain the legs and hypothenuse of a right angled triangle within a circle; also with the vertical leg outside the circle.

6. Explain, by the rule of three, the proportion between the sides and angles of triangles.   In this sines must be used as measures of degrees in working with degrees.

7. Illustrate the table of natural sines by a diagram.

8. Explain parallax generally.

9. Apply trigonometry to finding the moon's distance by its horizontal parallax.

10. Apply trigonometry to finding the sun's distance by the transit of Venus.

11. Apply the root and sines only in finding the height of a mountain, when the distance between the station and foot of the mountain is known, and angle at the base of the mountain between horizontal line and slant of hill.

12. Apply trigonometry to finding the length of a perpendicular of a right angled triangle, the base and sum of the perpendicular and hypothenuse being given.

13. Scale and dividers with all the lines on the scale.

14. Explain carpenter's sliding rule.

15. Explain sector and its use in perspective drawing.

16. Explain pantograph.

17. Explain spirit levels.

18. Glass thermometer and common ditto.

19. Explain barometer.

20. Hydrometer.

21. Explain hygrometer.

22. Explain quadrant, circular and quarter circle.

23. Explain sextant.

24. Pluviometer applied to rain and snow.

25. Compass, surveyors and navigators.

26. Chains and tallies, and why 9 stakes and 7 tallies are preferable.

27. Explain harbor surveying.

28. Illustrate the manner of working a traverse by sea or land.

29. Traverse about a field ; calculate the same by trapezoidal method.

30. Calculate the length of a degree of longitude at any degree of latitude.

31. Explain Mercator's chart.

32. Take the latitude of any place.

33. Take the longitude of any place.

34. Calculate the height of the atmosphere.

35. Calculate the pressure of the atmosphere upon any given surface on the earth by the barometer, say on a square yard.

36. Calculate the height of the lower valve of a pump at a given place by the barometer.

37. Cast the solid contents of a cone.

38. Cast the transverse diameter made by cutting an ellipse through the given frustrum of a cone.

39. Finish out a cone from a given frustrum.

40. Calculate a cask by assuming each end as a frustrum of a cone, without allowing for curvature.

41. Allowing for curvature, also the addition to the bung diameter of one tenth of the difference between bung and head.

42. Explain the method of calculating the angles of inflection in running a curve on a railroad when run on the periphery.

43. Explain the same when run by chord lines from one station.

44. Explain the method for calculating offsets from a chord line for fixing given equal points on a regular curve.

45. Show the method of calculating the quantity of water per second furnished by a running stream. Describe the best method for ascertaining the average velocity in a deep steam.

46. Illustrate contraction of the vein of water from an aperture.

47. Show that the velocity of effusions of apertures is increased as the square root of the height is increased ; taking 4 feet head giving 16.2 feet velocity per second, calculations may be made almost accurately.

48. Apply formula for determining the velocity and cubic feet of effusion per second under a given head.

49. Apply formula for determining the velocity and cubic feet under a given head through given cylinder waterworks.

50. Apply formula for calculating the velocity in open raceways and canals.

51. Apply formula for calculating the velocity and quantity of water pitching over a waste weir or dam.

52. Calculate excavations for canals.

53. Calculate embankments, dykes, docks, etc.

A few years later, in 1842, the following were given as " Qualifications requisite for a candidate for the degree of Civil Engineer :

He must be theoretically and practically familiar with trigonometry and mensuration, with their various applications.

He must be familiar with the level in laying out roads, M'Adam roads, railroads, canals, etc.

He must be perfectly familiar with running curves and staking out, and calculating for excavations and embankments.

He must be familiar with casting and constructing tables of ordinates and versed sines; also, the principles on which tables of natural sines are calculated, constructed and used.

He must be familiar with conic sections as far as they are used in civil engineering.

He must be familiar with statics and dynamics, and hydrostatics and hydrodynamics, so far as respects application to flumes, water-wheels, and descending raceways; also the velocity and efficient powers of spouting fluids applied to driving machinery.

He must be familiar by practice with the calculations for filling and emptying locks, the supply of water by weight and measure which any stream will afford as a feeder, or for any hydraulic purpose.

He must be familiar with taking the specific gravity of materials for construction.

He must be familiar with the necessary calculations for water-works ; whether conveyed in pipes, boxes or open raceways.

He must be familiar with calculating the height and pressure of the atmosphere.

He must be familiar with casting the height of clouds.

He must be familiar with taking and calculating latitude and longitude.

He must be familiar with taking the heights of hills and mountains with the barometer and thermometer ; also, with taking extemporaneous surveys and profiles with the barometer and triangular spans.

He must be qualified by practice to fix a transit line whenever required.

He must be qualified by practice to determine the variation of the needle at any time and place very nearly.

He must be qualified by practice to make a topographical survey of any district of country.

He must be qualified to change spherical areas of large districts, taken by latitude and longitude, into rectangular areas, by Mercator's method.

He must be an accurate land surveyor in theory and practice.

He must be a practical geologist, so far as to be able to make a correct report of the rocky and earthy deposits through which he lays out a canal or railroad.

He must be so far versed in architecture as to be enabled to direct the construction of bridges and other works of engineering in a comely style.

He must be perfectly familiar with plotting and business drafting."

CHAPTER VII.

THE fourth act of the Legislature relating to the Institute was passed May 8, 1837. It permitted the Troy Academy to be revived and united with the school. The new institution was to be named the Rensselaer Institute and was to consist of two separate branches, one to be called the department of experimental science and the other the department of classic literature. No such combination, however, resulted. By the same act the school was made subject to the visitation of the Regents of the University of the State and was declared to be entitled to the same privileges, government funds and other advantages as the academies, colleges and other schools of the higher order when it complied with the terms required by law and the rules of the Regents.

At a meeting of the trustees held September 25, 1841, the prudential committee was empowered to place the institution under the supervision of the Regents. Nothing was done in this direction, however, and on April 30, 1845, this committee was again authorized to consider the question. An application dated January 29, 1846, which contained a complete inventory and valuation of the property, was accordingly presented, and, in consequence, on the fifth of February of the same year the school was made subject to the visitation of the Regents, being classed as an academy until after its reorganization in 1849–50. Annual reports were made for eight years, and during this time it received a small amount of money, $744 in all, as its share of the literature moneys distributed to the academies of the state. In 1854 the authorities declined to make further reports, on the ground that the school had little in common with the academies. They were again made in 1869 and 1870, the institution being then classed as a scientific school. Another is found in the Report of the Regents for 1880, and since 1882 they have been made annually. They are now compulsory.

Upon the removal of the Institute, in May, 1834, from the Old Bank Place to the Van der Heyden mansion, a five-years' lease of the latter place was made ; and in order to provide proper facilities for the students the Patron caused a laboratory and study rooms to be built upon its grounds. After his death, which occurred January 26, 1839, the lease was renewed for two years. During this period the school suffered by the mutilation and final destruction, under the orders of the road commissioners of Troy, of the buildings erected by Mr. Van Rensselaer, and, as the agent of the property refused to restore them, at the expiration of the lease on May 1, 1841, a return to its original location was effected. Its second oc-

BUILDING ON INFANT-SCHOOL LOT. OCCUPIED, 1844-62.

cupation of the Old Bank Place was only three years in duration.

In 1843 the infant school lot situated on the northeast corner of State and Sixth streets, with a frontage of one hundred feet on Sixth Street and of ninety-eight feet on State Street, was offered as a gift by the city to the trustees, with the condition that William P. Van Rensselaer, a son of the founder, should give to the institution a sum of money equal to the value of the property. There was upon the lot a brick building fifty by thirty feet in size which was valued at $2500. The property was appraised at $6500, and, the condition being accepted by Mr. Van Rensselaer, was deeded to the trustees June 1, 1844. The $6500 in money thus obtained was invested as a permanent fund, and at the same time $1260 was raised by subscription for the purpose of building a laboratory. This was a one-storied brick building fifty by twenty-six feet in size, and was built upon the lot in 1844. It cost $1150. In the same year these two buildings were occupied by the school.

In the complete inventory contained in the application to the Regents made January, 1846, the buildings and lot were valued at $7650; the library of three hundred and ninety-six volumes at $973.45, and the surveying instruments, apparatus and specimens at $537.63. The money in possession of the trustees amounted to $6690, so that the total estimated value of the property of the Institution was $15,851.08. The total debts at the same time amounted to $1050.

In the catalogue for the thirty-fifth semi-annual session, published in 1841–2, during the second occupancy of the Old Bank Place, is given a list of students for the years 1839, 1840 and 1841, with their ages and addresses. During these three years there were seventy-seven students, most of whom came from the state of New York. Twelve of them, however, came from Connecticut, Maryland, New Hampshire, New Jersey, Pennsylvania, Tennessee, Vermont and Canada. Their ages varied generally between seventeen and twenty-five years, the average being twenty years. The list for the years 1840, 1841 and 1842, given in the catalogue of 1842–3, contains the names of seventy-five students, of whom ten were not residents of the state. One of them came from the territory of Wisconsin. During the next few years, until the extension of the course of study, the number varied between thirty-five and sixty-five annually, with an average age of about nineteen years. These numbers include students, of whom there was always a considerable number, who took partial courses and stayed only part of the year.

Amos Eaton having died May 6, 1842, George H. Cook, of the class of 1839, afterwards widely known for his work as State Geologist of New Jersey, was appointed Senior Professor and Agent September 19, 1842. He had previously been appointed Assistant Professor in March, 1840 ; Adjunct Professor of Civil Engineering in October, 1840, and Professor of Chemistry, Mineralogy and Zoology in September,

1841. His duties as Senior Professor included the delivery of courses of lectures on geology, chemistry and civil engineering. After somewhat extending the courses of study he resigned in 1846. His resignation was accepted by the board of trustees, with resolutions of regret, at a meeting held November 30, 1846, and on the same date B. Franklin Greene, Professor of Mathematics and Natural Philosophy in Washington College, Maryland, was appointed Senior Professor. He was graduated from the Institute in the class of 1842 with the degrees of Civil Engineer and Bachelor of Natural Science, and had been teaching at Washington College since 1843. In assuming the duties of Senior Professor he became at the same time Professor of Mathematics and Physics.

In the meanwhile the resignation of Dr. Nott had been accepted April 30, 1845, and Rev. Dr. N. S. S. Beman, who had been Vice-president since 1841, was elected President in his place.

The acceptance of the direction of the Institute by B. Franklin Greene marks an epoch in the history of the school. With the exceptions of its founder and Amos Eaton, it owes more to him than to any other person. Up to this date the course had been one year in duration, and although this length of time spent at the school did not necessarily insure the acquirement of either of the degrees, which were given only after satisfactory examinations had been passed, the average student who came reasonably well prepared could complete either of the courses in this period. After a careful study of the scientific and technical institutions of Europe Professor Greene thoroughly reorganized the curriculum. This reorganization, which included a material enlargement of the course of study and the requirement of a more rigid standard of scholarship from candidates for degrees, took place in the years 1849–50.

Professor Greene, who in the meanwhile had become Director of the institution when that office was created by act of Legislature in 1850, published in 1856 a pamphlet of eighty-seven pages, entitled "The Rensselaer Polytechnic Institute. Its Reorganization in 1849–50; Its Condition at the Present Time; Its Plans and Hopes for the Future." This, as its title indicates, was descriptive of the reorganization. Quotations from it will show more clearly the character of the changes and the intentions of the authorities :

"The managers of the Institute therefore resolved that *their field should be narrowed and more thoroughly cultivated;* that, indeed, their educational objects should be restricted to matters immediately cognate to Architecture and Engineering; that, moreover, for a somewhat irregular and for the most part optional course, requiring but a single year for its accomplishment, they would substitute a carefully considered curriculum which should require at the least three full years of systematic and thorough training ; and that, finally, they would demand the

application of the strictest examination tests to the
successive parts of the course prescribed, not only in
respect to the translation of students from lower to
higher classes, but, especially, in all cases of ultimate
graduation with professional degrees. It was in accordance with such views as these that, in 1849–50,
this institution was wholly reorganized upon the basis
of a general polytechnic institute, when it received
the distinctive addition to its title, under which it has
since been more or less generally known. Its objects
were thenceforward declared to be ' The education
of architects and civil, mining and topographical
engineers, upon an enlarged basis and with a liberal
development of mental and physical culture '."

" But it is proper to remark that, with the comprehensive statement and formal announcement, then
made, of what was proposed to be the future work of
the Institute, there was associated in the minds of its
managers no immediate expectation of realizing more
than a very partial development of their plans, with
the comparatively limited resources in *material* of
every kind at their command. Accordingly it was
resolved that, of the entire Institute curriculum, they
would at first proceed to develop the General Course
—the common scientific basis of the four professional
courses—and the two specialties of Civil and Topographical Engineering to as good a degree of excel-
lence as should be practicable under the existing circumstances ; while they would defer any attempt to
effect the more complete development of their plans,

including the important specialties of Architecture and Mining Engineering, to a period when they might hope to be able to invoke effectively the aid of conditions more favorable to realizations so desirable."

As indicated in these extracts, no attempt was made to develop at once all the special technical courses which it was intended to establish eventually. The course in Natural Science was made two years in length and that in Civil Engineering required three years. The first year was common to both. The degree given for the former course was Bachelor of Science, B.S., and for the latter Civil Engineer, C.E. The highest or senior class was called Division A and the others divisions B and C. In 1852 a " preparatory class ", in which students were fitted to enter Division C, was inaugurated.

An examination of the new curriculum shows the effect upon its formation of the study of the French scientific schools. Its object was practically that of L'École Centrale des Arts et Manufactures, which, in a three-years' course, was intended to train civil engineers, directors of works, superintendents of manufactories, professors of applied science, etc., and the reorganized course bears considerable resemblance to that of the same school. That part of it which forms the groundwork for the higher technical studies also resembles the curriculum of L'École Polytechnique, which, it will be remembered, does not furnish a complete system of instruction, but has for its object

IN THE DYNAMO ROOM. 1894.

the preparation of students for entrance to certain government technical institutions.

It was the intention to obtain, as far as the conditions would admit, the same end here in a single school that was obtained in France from L'École Polytechnique and the special schools combined. As a matter of fact, with the same high aim in view, the curriculums of such institutions, wherever situated, must necessarily bear a resemblance to each other. In relation to this subject the circular of February, 1851, informs us that " In the essential features of its design and intentions the Institute may be said to occupy a position between L'École Polytechnique and L'École Centrale des Arts et Manufactures, of Paris. It claims no other resemblance to these celebrated and *richly endowed* institutions. To its peculiar mode of study *there is no known counterpart.*"

The mode of study at this time contained the essential features of that which characterized the beginnings of the school. The students took full notes of the lectures delivered by the professors and afterwards studied the subjects by the aid of their notes, their own practical exercises and books of reference. The next day they were interrogated by the instructors and after the interrogation were divided into small sections which assembled in different rooms. Each student then delivered an extemporaneous lecture upon the subject under consideration, which was afterwards criticised by the other members of his section and by an officer styled a " Repeater ", who, under the direction of the professor at the head of the department, took charge of the several sections.

The Repeaters were generally resident graduates or students who were members of the highest class in the institution. The term seems to have been taken from the name *Répétiteur*, given in L'École Centrale to a class of instructors with similar duties. It was used only a few years and appears for the last time in the catalogue of 1859, in which, among twelve instructors, there is found only one, the Repeater of Mechanics, who was at the same time Assistant Professor of Mathematics. In that of 1855, among eleven instructors there is no repeater. The practice of requiring daily lectures from each student was gradually dropped with the use of this title, and the present method of strict interrogations and of blackboard demonstrations which partake of the nature of the lectures, was as gradually introduced. This change was largely and almost necessarily the result of the increased attendance at the school.

The " Notices " of 1835 and the examination questions of the succeeding year, together with the qualifications required of candidates for degrees in 1842, all of which are found in the preceding chapter, give a reasonable knowledge of the character of the work done at that period of the school's history. As it is now proposed to set forth the curriculum after the reorganization, it will be well to preface it with the remark that, although the limited time given to the course naturally restricted its value, gradual improvements had been made in the intermediate years, as required by the advances in natural and applied science. In fact, the reorganization itself was not immediately completed. Although it may be said to have taken place in 1849–50, and the courses were extended at this time, a departure, in most respects so decided, from its previous methods necessarily could not be immediately accomplished. By the year 1854 the courses in Civil Engineering and Natural Science had been well developed. The table which follows, taken from the Annual Register of that year, gives an outline of the subjects studied and the order of their distribution.

SCHEDULE OF THE COURSE IN CIVIL ENGINEERING (1854).

Departments of Instruction.   Subjects of Study.

FIRST YEAR.

FIRST TERM.

Mathematics............. Algebra—Geometry.
General Physics......... Molecular Forces—Thermotics.
Graphics................ Geometrical Drawing: *Elementary Drawing.*
Geodesy................. Line Surveying: *Theory* (Commenced); *Field Work.*
English Composition..... The Course (Commenced).
French Language......... The Course: *French Grammar.*

SECOND TERM.

Mathematics............. Trigonometry—Higher Algebra.
General Chemistry....... Non-metallic Chemistry.
Graphics................ Topographical Drawing; *General Topography; Maps of Farm Surveys.*
Geodesy................. Line Surveying: *Theory* (Finished); *Office Work.*

Natural History.......... Botany.
English Composition..... The Course (Continued).
French Language........ The Course: *Translations from French into English.*

SECOND YEAR.

First Term.

Mathematics............ Analytical Geometry—Differential Calculus.
General Physics......... Electricity.
General Chemistry....... Metallic Chemistry.
Natural History......... Mineralogy.
Graphics .... Descriptive Geometry: *General Theory*—Geometrical Drawing: *Architectural Drawing.*
Geodesy................ Practical Trigomometry.
English Composition..... The Course (Continued).
French Language........ The Course: *Reading from French Scientific Authors.*
German Language........ The Course: *German Grammar.*

Second Term.

Mathematics............ Integral Calculus.
General Physics.......... Acoustics—Optics.
Natural History.......... Zoology.
Geology and Physical
  Geography......... Geology.
Graphics..... .... Descriptive Geometry: *Shades and Shadows*—Geometrical Drawing: *Machine Drawing.*
Geodesy................ Topographical Surveying—Hydrographical Surveying.
English Composition..... The Course (Continued).
German Language........ The Course : *Translations from German into English.*

THIRD YEAR.

First Term.

Mechanics.............. Mechanics of Solids — Mechanics of Fluids.
Practical Astronomy...... The Course (Commenced).
Physical Geography...... The Course.
Practical Geology....... The Course.
Geodesy................ Trigonometrical Surveying.

Graphics................. Descriptive Geometry: *Perspective*; *Isometrical Projection* — Topographical Drawing: *Maps of Trigonometrical Surveys*.

Machines................. Theory of Machines.

Industrial Physics........ Practical Pneumatics —Practical Thermotics.

Philosophy of Mind...... The Course (Commenced).

English Composition..... The Course (Finished).

SECOND TERM.

Constructions............ Theory of Structures —General Constructions — B r i d g e s — Hydraulic Works—Railways.

Machines................. Prime Movers—Special Machines.

Mining .... The Course.

Practical Astronomy...... The Course (Finished).

Geodesy................. Railway Surveying—Mine Surveying.

Graphics................. Descriptive Geometry: *Stone Cutting—* Topographical Drawing: *Maps, etc., of Railway Surveys*; *Plans, etc., of Mine Surveys*.

Metallurgy .... General Metallurgy — Metallurgy of Iron.

Industrial Physics........ Architectural Physics.

Philosophy of Mind...... The Course (Finished).

SCHEDULE OF THE COURSE IN NATURAL SCIENCE.

FIRST YEAR.

The course for the first year is the same as that in Civil Engineering.

SECOND YEAR.

FIRST TERM.

General Physics.......... Electricity.

General Chemistry....... Metallic Chemistry.

Natural History.......... Mineralogy.

Geology and Physical Geography............. Physical Geography.

Practical Geology........ The Course.

Graphics................. Geometrical Drawing: *Architectural.*

Industrial Physics .... Practical Pneumatics — Practical Thermotics.

Philosophy of Mind......	The Course (Commenced).
English Composition.....	The Course (Finished).
French Language........	The Course: *Reading French Scientific Authors.*
German Language........	The Course: *German Grammar.*

SECOND TERM.

Natural History..........	Zoology.
Geology and Physical Geography	Geology.
General Chemistry........	Organic Chemistry.
Natural History Applied to the Arts.............	The Course.
General Physics..........	Acoustics—Optics.
Industrial Physics.......	Architectural Physics.
Philosophy of Mind......	The Course (Finished).
German Language.......	The Course: *Translations from German into English.*

In the table the use of the term "The Course" after a subject refers to a detailed description of it in an exhaustive schedule which follows the table in the Register. This gives in minute detail the scope of each subject taught and the text-books and works of reference used. It covers forty pages, containing thirty-one main and two hundred and two sub-divisions.

Lectures and text-books were both used in most of the courses. Among the text-books may be mentioned : Davies' Legendre's Geometry, Davies' Bourdon's Algebra, Chauvenet's Trigonometry, Church's Analytical Geometry, Church's Calculus, Mahan's Industrial Drawing, Davies' Shades, Shadows and Perspective ; Davies' Descriptive Geometry, Jopling's Isometrical Perspective, Davies' Surveying, Simms' Mathematical Instruments, Gummere's As-

A CORNER IN THE PHYSICAL LABORATORY.   1894.

tronomy, Hitchcock's Geology, Dana's Mineralogy, Gray's Botany, Gregory's Elements of Chemistry, Mill's Qualitative Analysis, Fresenius' Quantitative Analysis, Morfit's Chemical Manipulation, Bird's Natural Philosophy, Bartlett's Acoustics and Optics, Bartlett's Analytical Mechanics, Weisbach's Mechanics of Machinery and Engineering, Pambour's Theory of the Steam Engine, Moseley's Mechanical Principles of Engineering and Architecture, Morin's Aide-Mémoire de Mecanique Pratique, Haupt's Bridge Construction, Mahan's Civil Engineering and D'Aubuisson's Traité d'Hydraulique. A list of one hundred and twenty-nine works of reference in English, French and German is also given.

The practical part of the work of the school included surveys, chemical and physical laboratory work, botanical and geological excursions, visits to factories, etc.

Applicants for admission were required to be at least sixteen years old. The majority were over eighteen. They were required to be well prepared in geography, English composition, arithmetic, including the metric system ; plane geometry, and algebra to equations of the second degree.

The first "Register" to appear after the reorganization was a pamphlet of sixteen pages dated August 15, 1851. The second, which was published in October, 1852, contained after the names of the students their grades in the different departments and their class standing. After some of them the letters "d" and "a", meaning respectively "deficient" and "not examined", were placed. To this there was decided objection on the part of the students, who republished this register in December of the same year, leaving out the objectionable features. The grades were in consequence omitted from succeeding registers, though the "order in general standing" upon graduation was published until 1855, since which year all names of undergraduates have appeared in alphabetical order in the different divisions.

About this time students were advised to wear a "uniform dress", and many of them did so. The suit, including a cap, was made of dark-green cloth. The coat was a single-breasted frock with a black velvet collar, and the cap had an ornamental symbol in gold placed on the band in front. The custom did not continue very long, and the uniform was officially mentioned for the last time in the Register of 1855.

Shortly after the extension of the course of study the name of the school was changed from the Rensselaer Institute to the Rensselaer Polytechnic Institute. In a "Programme" issued in 1851 it is called by its former name, but in the Register published in August of the same year the latter title is used. Although henceforth known as the Rensselaer Polytechnic Institute, the change was not ratified by act of Legislature until April 8, 1861. The name "Annual Register" was first given to the official catalogue in 1854.

The improvement of the curriculum was followed by an increase in the number of students and instructors. The report to the Regents of the University of the State, made in 1848, shows that on September 29 of that year there were twenty-two students, and that during the year ending on that date there had been a total attendance of fifty-one. The number of instructors was five, including the president, who lectured once a week on Mental and Moral Philosophy. In 1855 there were one hundred and fourteen students, of whom fifty-one were from the state of New York, forty-eight from fourteen other states, including Maine, Louisiana and California, and fifteen from foreign countries. The number of instructors had increased to eleven, including Dr. Beman. In consequence of the extension of the course no class was graduated in 1852.

In 1848 the tuition was $20 for each term of five months, or $40 a year. Those who worked in the chemical laboratory paid $8 a term more. In 1851 the corresponding fees were $60 a year and $5 a term. In 1857 the tuition was $100 a year, with no extra charges. This was increased to $150 a year in 1864 and again in 1866 to $200, at which price it still remains.

The fifth act, relating to the institution, passed by the Legislature of the state was dated March 8, 1850. Beside creating the office of Director this law reorganized the board of trustees. It was enlarged to nineteen members, and the only *ex-officio* member left in it was the Mayor of Troy. All restrictions as to place of residence of members were abolished. The act of April 8, 1861, which legalized the change of name of the Institute made ten years before, also gave the board the power to increase its number to twenty-five members, including the Mayor of Troy. No further change has since been made in this number. By the same law the Trustees were given the power to confer the degrees of Civil Engineer, Topographical Engineer, Bachelor of Science and such other academic honors as they might see fit. This was merely a more explicit definition of their power to grant certificates than was given by the act of 1835, under which they had been annually conferring degrees.

In pursuance of the plan outlined at the time of the reorganization a course in Topographical Engineering was, in 1857, added to those already existing. Upon its satisfactory completion the candidate received the degree of Topographical Engineer, T.E. Like the course in Natural, or, as it was then called, General Science, it was two years in length, while that in Civil Engineering required three years. A special course in Land Surveying, only one year in duration, was also inaugurated. The first year of the Topographical curriculum was identical with that in Civil Engineering. In the second year pure mathematics, graphics, physics, chemistry and geology were taught, and especial attention was given to

general surveying, practical astronomy and topographical drawing.

It will be remembered, in considering the time
given to the three principal courses, that the preparatory class increased their length for some of the students by a period of one year. Since the first year
of its establishment its members had varied in number from twenty-two to thirty-two. They were
treated as members of the Institute, and their names
were printed in the Register, after Division C, under
the heading " Preparatory Class ". In 1858 " Division D " was prefixed to this title, and after 1862 it
was no longer called the preparatory class but simply
" Division D ".

In 1860 the special course in Land Surveying was
abolished and the courses in General Science and
Topographical Engineering were made three years
in length, the same as that in Civil Engineering. In
1862, when the preparatory class became Division D,
the latter course was made four years in length and
the two former each three years. These two, however, began with Division C, the course in Topographical Engineering being identical with that in
Civil Engineering throughout the work of divisions
C and B, and the course in General Science coincid-
ing with both of the engineering courses in Division C.

At this time candidates for admission to Division
D were required to be not less than fifteen years old,
and they were examined in geography, English

grammar, arithmetic and algebra (through equations of the first degree).)

During the scholastic year 1862–3 still other changes were made, a course in Mechanical Engineering was added, and each of the four courses was made four years in length, the first two years being identical in all. The last two years in Mechanical Engineering contained, of course, more of the theory and practice of machine construction than those leading to the other two professional degrees. Courses in Structures and Hydraulics were more largely developed in the Civil Engineering curriculum, and Geodesy and General Surveying in that of Topographical Engineering. The improvements in these various courses, made annually during the preceding years, are given in detail in the Annual Registers.

In 1866 the course in Topographical Engineering was replaced by one in Mining Engineering. The number of students in the former had never been great, and of these only five had been graduated, all in the class of 1860. The first two years in Mining Engineering were identical with those of the other courses. The distribution of the subjects in the last two years will be given here.

TRIGONOMETRIC SURVEYING.   1894.

SCHEDULE OF THE TWO LAST YEARS OF THE COURSE IN MINING ENGINEERING (1866).

Departments of Instruction.   Subjects of Study.

DIVISION B.

FIRST TERM.

Mathematics............. Differential Calculus—Integral Calculus —Method of Least Squares.

Physics................. Electricity: *Terrestrial Magnetism; Statical and Dynamical Electricity.*

Chemistry............... Qualitative Analysis: *Behavior of bases and acids with reagents.*

Natural History.......... Mineralogy.

German Language........ German Grammar — English Translations.

Geodesy................ Practical Trigonometry — Levelling — Topographical Surveying.

Geometrical Drawing..... Machine Drawing: *Elements of Machines.*

Topographical Drawing .. Maps of Topographical Surveys.

SECOND TERM.

Rational Mechanics ...... Mechanics of Solids — Mechanics of Fluids.

Descriptive Geometry.... Linear Perspective.

Physics................. Acoustics and Optics.

Chemistry............... Qualitative Analysis.

Natural History.......... Mineralogy— Geology—Zoology—Palæontology.

German Language........ English Translations.

Geometrical Drawing..... Perspective.

Topographical Drawing.. Colored Topography.

DIVISION A.

FIRST TERM.

Physical Mechanics ...... Mechanics of Solids: *Friction; Strength of Materials.* Mechanics of Fluids: *Practical Hydraulics; Practical Pneumatics.*

Machines................ Theory of Machines.

Descriptive Geometry.... Stone Cutting.

Chemistry............... Qualitative Analysis—Metallurgy.

Natural History..........	Mineralogy—Geology.
Philosophy	Intellectual Philosophy.
Geometrical Drawing.....	Stone Cutting.

SECOND TERM.

Machines................	Theory of Prime Movers: *Steam Engine.* Designs for and Reviews of Special Machines.
Chemistry...............	Quantitative Analysis—Metallurgy—Assaying.
Geodesy.................	Mine Surveying.
Practical Mining........	Sinking and Driving—Ventilation and Drainage—General Management.
Philosophy	Ethical Philosophy.

In July, 1859, B. Franklin Greene severed his connection with the Institute, after a service of more than twelve years. At first Senior Professor with the chair of Mathematics and Physics, his title was changed in 1850 to Director and Professor of Physics, Chemistry and Geology. In 1852 he became Professor of Physics, Mechanics and Constructive Engineering, and in 1855, Professor of Mechanics, Machines and Constructions. The change in the character of the course while he was at the head of the faculty gives evidence of his efficiency and great ability.

Ever since he had been elected Vice-president in 1841, Rev. Dr. Beman had delivered lectures on Mental and Moral Philosophy at the Institute, and since 1854 he had been Professor of Mental Philosophy as well as President of the Board of Trustees. Upon the resignation of B. Franklin Greene he was made Director as well, and the title of Senior Professor was revived and conferred upon Charles

Drowne, who became at the same time Professor of Civil Engineering. Professor Drowne was graduated in the class of 1847 with the degree of Civil Engineer, and in the same year became Assistant in Mathematics and Physics. In 1850 he was Adjunct Professor of Theoretical and Practical Mechanics, and from 1851 to 1855 Professor of Mathematics, Astronomy and Geodesy. Dr. Beman remained Director only one year, and in 1860 Charles Drowne became Director and Professor of Theoretical and Practical Mechanics. The term Senior Professor was then dropped and has not since been used.

Although resigning as Director, Dr. Beman continued President of the Board of Trustees until advancing years compelled him to terminate, in 1865, his long and useful connection with it. He was succeeded, March 20, 1865, by John F. Winslow, one of the proprietors of the Rensselaer Iron Works of Troy. He had been a trustee since 1860. Mr. Winslow retained his position only three years; his removal to Poughkeepsie causing him to resign April 9, 1868. On May 7 of the same year the sixth President, Dr. Thomas C. Brinsmade, was elected. He was a physician of Troy who had been a trustee for twenty-four years, having been elected March 4, 1844, during the second occupation of the Old Bank Place. His term of office was short. Whilst reading a paper on the condition of the Institute at a public meeting, held in the evening of June 22, 1868, for the purpose of raising funds for the school, he died suddenly of heart disease. James Forsyth, a lawyer of Troy, was made President December 15, 1868. He had not previously been connected with the institution.

.

CHAPTER VIII.

PRESENT EQUIPMENT—MISCELLANEOUS INFORMATION.

A GREAT fire which swept over many blocks and destroyed property valued at nearly three millions of dollars occurred in the city on May 10, 1862. It burned the buildings of the Institute, which, beside the two already described, included one adjacent to them obtained shortly before the fire for a mineralogical and geological museum. The furniture, geological specimens and a part of the chemical apparatus were also destroyed, though a portion of the apparatus and the library were saved.

Temporary quarters were immediately obtained in the University Building on the hill, now called the Provincial Seminary, and the course was resumed on the following Wednesday. Accommodations for the next year were secured in the Vail Building, on the northeast corner of Congress and River streets; and the school remained there until the completion, in May, 1864, of the structure on Eighth Street, at the head of Broadway, which, under the name of the Main Building, is still used for purposes of instruction. It is built of brick and is one hundred and fifteen feet long by fifty feet wide, consisting of a central portion five stories in height and two wings, each of four stories. The land upon which it is situated, including that now occupied by the Winslow Laboratory, was given by the Warren family of Troy, Joseph M. Warren, one of its members, having been a trustee and firm friend of the school since 1849.

The construction of a chemical laboratory was begun in 1865 on that part of the grounds north of the Main Building. It was named the Winslow Laboratory, in honor of President John F. Winslow. He had always been deeply interested in the prosperity of the school, and had contributed largely toward the construction of the Main Building. The laboratory, which was completed during the summer of 1866, was built of brick and was sixty feet long by forty feet wide, and three stories in height. During the night of August 27, 1884, the upper story, containing lecture and recitation rooms and the chemical library, was burned, and much apparatus and nearly a thousand volumes were lost. It was rebuilt and ready for occupancy by February, 1885. In the rebuilding it was improved and enlarged, and is now seventy-three feet long by forty feet wide, and three stories in height.

In 1871 it was determined to improve the course in Civil Engineering and concentrate the efforts of the school upon it. The three courses in Natural Science, Mechanical Engineering and Mining Engineering were therefore abolished. The number of

HYDROGRAPHICAL SURVEY. 1894.

students taking the first two had been small, and, although more had taken the last, between the years 1868 and 1871 only twenty-three had been graduated with the degree of Mining Engineer. Metallurgy and free-hand drawing were added to the civil engineering curriculum, and the courses in chemistry, physics and geology, as well as those in a number of the practical engineering subjects, were extended and improved. In the course as developed a wide significance was given to the term civil engineering, as is shown by the inclusion in the course of such subjects as metallurgy, thermodynamics, the theory and construction of engines and other machines, etc.

There was at this time, as there always has been, a considerable number of students who took special courses and were not candidates for a degree. After a lapse of fourteen years the course in Natural Science was re-established at a meeting of the trustees held September 23, 1885, and still continues a department of instruction at the Institute.

The semi-centennial celebration of the foundation of the school was held at Troy, June 14 to 18, 1874. Besides the usual commencement exercises there was a largely attended alumni meeting, three days in duration, at which historical and other addresses pertinent to the occasion were made by the President, graduates, professors and others. A monument to Amos Eaton, which had recently been placed in Oakwood Cemetery, was dedicated, and sketches were given of the lives of five graduates and students

who had served in the civil war and for whom me-
morial windows had recently been placed in the Main
Building. These were Major James Cromwell, C.E.,
Colonel Charles Osborn Gray, Major Otis Fisher,
Lieutenant Henry W. Merian, C.E., and Major
Albert Metcalf Harper, C.E. Shortly after the
meeting a sixth window, to the memory of Captain
James R. Percy, C.E., was added. These six memo-
rials, however, did not represent all of the graduates
and students who had been in the war. More than
seventy-five had served in the army and navy of the
United States, in various capacities, during that period.

In 1874 memorial windows to Amos Eaton and to
Professors John Wright and William Elderhorst were
also placed in the assembly hall of the Main Building.
Professor Wright had held the chair of Botany and
Zoology from 1838 to 1845, and William Elderhorst
had been Professor of Chemistry from 1855 to 1861.

A leave of absence was granted Professor Drowne,
in November, 1875, on account of ill health. He did
not recover sufficiently to enable him to return, but
resigned December 9, 1876, on which date William
L. Adams was appointed Director. President Forsyth had been acting in this capacity from December 11, 1875, until his appointment. Professor
Adams was a graduate of the class of 1862. After
some experience in the field he became Acting
Professor of Geodesy, Road Engineering and Topographical Drawing from September, 1864, to February, 1865, when he resumed the active practice of his

profession. In September, 1872, he returned to the Institute to take charge of the department in which he had previously been Acting Professor. He again left, in 1878, to return to the profession of railroad engineering, and on September 10 of the same year David M. Greene of the class of 1851 was elected Director. Professor Greene had been for a short time after his graduation Assistant in Mechanics and Physics at the Institute, and had occupied the chair of Geodesy and Topographical Drawing from 1855 to 1861.

The third building to be erected for purposes of instruction was an astronomical observatory which was finished in 1878. It was presented by Mr. and Mrs. Ebenezer Proudfit of Troy as a memorial to their son Williams Proudfit, a bright and promising student of the class of 1877, who was, in 1875, fatally injured by being thrown from his carriage. The trustees received a letter from the donors November 6, 1875, in which they signified their intention to erect the observatory. In consequence, a suitable site was found in the Ranken property, situated on the east side of Eighth Street, nearly opposite to the chemical laboratory. This was bought by the Board January 25, 1877. It has a frontage of one hundred and fifty feet on Eighth Street and extends eastward about five hundred feet to the brow of a hill which has an elevation of about two hundred feet above the Hudson River. The property included a dwelling-house and stable, both built of brick. The house forty feet square and two stories in height, now contains the testing-machines of the school.

The Williams Proudfit Observatory, situated on the brow of the hill, is built of brick with stone trimmings, and consists of a central part thirty feet square, with three wings, the total length being seventy-six feet and breadth sixty feet. The main part is two stories high, with a dome twenty-nine feet in diameter, under which is the main pier intended for an equatorial telescope. The wings are each one story in height, that to the east containing the transit instrument and other apparatus used for astronomical purposes.

During the alumni meeting held at Troy in June, 1881, a committee of graduates was appointed to solicit funds for the endowment of the institution. Francis Collingwood, '55, was made chairman, and the other members were : George W. Plympton '47, William H. Martin '56, William Metcalf '58, Joseph M. Wilson '58, Robert Neilson '61, Arba R. Haddock '62, Frederic W. Vaughan '63, Joseph C. Platt '66, Joseph W. Campbell '68, Thomas Appleton '68, Theodore Voorhees '69, Arthur E. Boardman '70, David Reeves '72, Frank L. Rowland '75, J. F. Aldrich '77, Conrad B. Krause '79, and George A. Just '81.

This action was approved at the meeting held in New York City in January, 1882, and was officially sanctioned by the board of trustees February 24, 1882. On this date the board appointed James P.

Wallace '37, E. Thompson Gale '37, and Charles Macdonald '57, as a committee to receive and manage the funds, which were to be deposited, pending investment, with the Central Trust Company of New York City. It was concluded to make an effort to raise $100,000 for the purpose of endowing the Directorship. Subscriptions were solicited from graduates, and although the whole sum has not yet been raised, partly because other demands have since been made upon them, notably for the construction of the Alumni Building, a considerable proportion has been received. The amount collected was largely due to the efforts of Mr. Collingwood.

The year 1883 is made memorable by the endowment of the chair of Rational and Technical Mechanics. Sixty thousand dollars was given for this purpose by Mrs. Mary Elizabeth Hart, as a memorial to her husband, with the condition that the chair should be designated the William Howard Hart Professorship of Rational and Technical Mechanics. The communication to the board of trustees offering the endowment was dated June 11, 1883, Mr. Hart having died on the third day of the preceding April. He was the son of Richard P. Hart, who had been a trustee of the school in its earlier days (1825–43), and was a man of fine character ; an earnest student of nature with strong scientific tastes. He had always been interested in the school, and in her letter Mrs. Hart informed the board that the endowment was " in furtherance of his views and as a fitting memorial of his interest in the prosperity and success of the Institute ". It is proper to remember, nevertheless, that the gift was due to Mrs. Hart and was an evidence of that benevolence of character which has since been shown in so many ways, to the benefit of her native city.

In May, 1883, a petition was received by the trustees from the students, who asked that steps be taken by the board to provide a suitable gymnasium for their use. The subject was again agitated later in the year, and in 1884 a lot on the south side of Broadway, at the foot of the property containing the Main Builing, was purchased by the trustees. Upon this site a gymnasium of brick, trimmed with stone and terra-cotta, eighty feet long by forty-four feet wide, and two stories in height, was erected. It was opened March 11, 1887. About half the money expended in its construction was contributed by alumni, trustees, students and residents of Troy, and the remainder was appropriated from the funds of the institution. The first story contains a reception-room, a dressing-room, shower-baths and bowling-alleys, and the second the main hall, which is about thirty feet high and is fitted with the best patterns of gymnastic apparatus. There is a running track around this hall and at one end a gallery for spectators.

It has been seen that the geological and mineralogical specimens belonging to the school were destroyed by the fire of 1862. Another collection was immediately begun by Professor H. B. Nason,

PRELIMINARY SURVEY FOR A RAILROAD. 1894.

at that time Professor of Natural History, who was then in Europe. A thousand dollars was given for this purpose, and by the fall of 1862 more than a thousand specimens of minerals, rocks and fossils had been obtained. Since then their number has been constantly augmented and collections in other branches of natural history have been accumulated. At the present time the cabinets of minerals, rocks, fossils, etc., contain more than ten thousand specimens, the collection of shells numbers about seven thousand, of birds about four hundred, of specimens of wood nearly three hundred, and of plants about five thousand.

The library, composed almost wholly of scientific books, has also been constantly increasing in value. It consists at present of about six thousand volumes and about three thousand pamphlets and maps. It contains many engineering works, including the publications of foreign and American scientific societies, and bound volumes of all the more important technical journals. The valuable professional library and drawings of Alexander L. Holley, formerly a trustee of the Institute, was bequeathed to it in 1882.

For many years the geological collections and cabinets of natural history were kept in a large hall on the top floor of the Main Building and the library was in a room on the second floor. The erection of a fire-proof building in which both could be safely kept was urged by Professor Nason at the Alumni meeting in Troy, June 13, 1888. The State Geologist of New York, Professor James Hall of the class of 1832, had promised to give a valuable collection of fossils if such a building were provided. Part of the amount required for its construction was raised by subscription from graduates at the meeting, and at the Pittsburgh meeting of the association of graduates held January 31, to February 1, 1889, enough was pledged to insure its erection. A lot on the east side of Second Street, between State Street and Broadway, immediately north of the Savings Bank building, was purchased June 2, 1890, with a fund raised by subscription among the trustees, and the building was completed in 1893. Wilson Brothers and Co. of Philadelphia provided the plans, the three brothers from whom the firm takes its name being graduates of the Institute. The structure is fireproof, fifty feet square and three stories in height. The lower portion is faced with brownstone and the upper with yellow brick and terra-cotta. The library, a room for the trustees and the office of the Director are on the first floor, and the other two contain the geological, mineralogical and general natural history collections. There is also a lecture-room for the department of Geology on the second floor.

President Forsyth, who beside his official duties as President of the Board of Trustees had lectured on the Law of Contracts since 1873, died August 10, 1886. Upon his death, William Gurley of the class of 1839, the Vice-president of the board, became Acting President and remained so until his death

January 11, 1887. On June 1 of the same year Albert E. Powers, a banker and manufacturer of Lansingburg, who had been a trustee since 1861, was elected Vice-president and acted as President until May 2, 1888, when John H. Peck, a prominent lawyer of Troy, was elected to that office. Mr. Peck had been a member of the board of trustees since June 1, 1887. He is still (1895) President and is also lecturer on the Law of Contracts.

After a service of thirteen years David M. Greene resigned September 15, 1891, and Professor Dascom Greene, at the head of the department of Mathematics and Astronomy, was appointed temporary Director. He held this position until the election, January 15, 1892, of Palmer C. Ricketts of the class of 1875, who had been Assistant in Mathematics and Astronomy from that year until 1882 and Assistant Professor in the same department from 1882 until 1885, when he became William Howard Hart Professor of Rational and Technical Mechanics. He is still (1895) Director and Professor of Mechanics.

Ever since the reorganization of the Institute by B. Franklin Greene each candidate for a degree has been required to present a thesis on some subject germane to his course. Such theses are read at commencement, and one of the conditions for graduation is that they must be approved by the faculty. In order to improve their quality Charles Macdonald, C.E., LL.D., of the Union Bridge Company, a graduate of the class of 1857, established, September 24,

1890, a prize consisting of the net annual income from $2000, to be given to that member of Division A, in each year, who should, on graduating, present the best thesis involving a design for an engineering work or an investigation of a process or natural product, or of a natural law of especial interest to civil engineers. This prize is awarded at the commencement following that at which the competitor graduates. It has already proved of much value as it increases the interest taken in their theses by those students competing for it, and incidentally has been effective in improving the character of all which are presented. Three graduates have received it : from the class of 1891, Stacey E. Denny ; from that of 1892, Elmer J. Bucknell ; and from the class of 1893, Ralph H. Chambers. The successful competitor in the class of 1894 is Paul L. Reed of Denver, Colorado.

The Alumni Association of the Institute was organized at Troy June 22, 1869. Annual meetings are held on commencement day of each year at Troy, and of late years it has been customary to hold winter reunions some time during February in one of the larger cities of the country containing a considerable number of resident graduates. Such meetings have been held in New York, Philadelphia, Pittsburgh, Buffalo, Kansas City and Cleveland, and a summer meeting was held, during the Columbian Exposition, at Chicago in August, 1893. The first general reunion was held February 18, 1881, in New York, at the residence of Hon. Clarkson N. Potter of the class of 1843. The names of graduates who have been presidents of the Association, with their terms of office, are as follows : James Hall, 1869–71 ; Albert R. Fox, 1871–73 ; Strickland Kneass, 1873–74 ; William Gurley, 1874–78 ; John G. Ambler, 1878–79 ; James P. Wallace, 1879–80 ; Francis Collingwood, 1880–81 ; Charles Macdonald, 1881–83 ; Charles C. Martin, 1883–84 ; Joseph M. Wilson, 1884–85 ; Joseph C. Platt, 1885–86 ; David Reeves, 1886–87 ; Theordore Voorhees, 1887–88 ; T. Guilford Smith, 1888–89 ; Christopher C. Waite, 1889–90 ; Joseph J. Albright, 1890–91 ; Clark Fisher, 1891–92 ; William B. Cogswell, 1892–93 ; Theodore N. Ely, 1893–94, and William Metcalf, 1894.

Several local alumni associations have recently been formed ; one, February 10, 1888, at Kansas City called the "Central R. P. I. Association"; the "Pittsburgh Association of Graduates" May 11, 1888 ; the "Chicago R. P. I. Association" November 25, 1889, and the "R. P. I. Alumni Association of New York City" January 17, 1893.

In past years a number of attempts were made by undergraduates to publish periodicals in the interest of the students and alumni of the school. The first number of the *Rod and Leveller* appeared November 18, 1865 ; and in May, 1884, the *Rensselaer Polytechnic Institute Quarterly* was issued for the first time. These failed shortly after their inception. A successful effort in this direction, however, was made by Tracy C. Drake of the class of 1886, and the first number of the *Polytechnic*, with him and A. R. Elliott as editors, appeared February 16, 1885. Since that time it has been issued regularly each month during the scholastic year and is now well supported by students and alumni. It is published by a board of editors from different classes and each issue contains about twenty-five quarto pages of scientific and literary articles and of news items relating to the school and its graduates.

The *Transit*, an annual issued under the auspices of the Fraternities by a board of editors selected from members of Division B, has been published for twenty-nine consecutive years. The first number, dated December, 1865, was issued by the class of 1867. Beside the roll of members of the classes, fraternities and societies it contains lists of members of the athletic, glee and other clubs and miscellaneous organizations.

The "Selected Papers" of the Rensselaer Society of Engineers are also published at irregular intervals. These are often of much scientific value.

The first college fraternity to establish a chapter at the Institute was the Theta Delta Chi. The Delta chapter was chartered in 1853 and remained until 1870. It was re-established in 1883. Beside this there are five others existing at present: the Alpha chapter of Theta Xi (1864), Lambda of Delta Phi (1864), Psi Omega of Delta Kappa Epsilon (1867), Theta of Chi Phi (1878) and Upsilon of Delta Tau Delta (1879). The Pi chapter of Zeta Psi was

TESTING CEMENT. 1894.

established in 1865 and withdrawn in 1893. Several others were chartered at various times but were withdrawn after an existence of one or two years.

The Pi Eta Scientific Society, organized January, 1866, became afterwards the Rensselaer Society of Engineers, which was incorporated by act of legislature in May, 1873. Papers are read by the student members at the meetings throughout the year and scientific lectures are also delivered at intervals by graduate members of the society and others.

The Zeta chapter of the Sigma Xi Society was established at the Institute May 6, 1887. This society is modelled to some extent after Phi Beta Kappa, though it is not a secret society. Its undergraduate members are chosen only from those who have distinguished themselves in scholastic work.

The Institute has had exhibits at three world's fairs. It sent some students' drawings to the World's Industrial and Cotton Centennial Exposition, held at New Orleans in 1884–85, and received a medal and diploma of the "First Order of Merit" for mechanical and free-hand drawing. It also obtained for its exhibit at the Universal Exposition of the French Republic at Paris, in 1889, the only grand prize awarded to any American scientific school. At the World's Columbian Exposition of 1893, in Chicago, it exhibited the work of its students and graduates and received awards for each, worded as follows: "Superior instruction in matter and method, through its long continued service.

Marked attainments of its students in all forms of class work, including topography, railroad maps, mechanical drawing and theses"; and "The magnificent work of its graduates, including (*a*) the arches of the Liberal Arts building, (*b*) the Ferris Wheel, (*c*) the Brooklyn Bridge, (*d*) the Poughkeepsie Bridge, (*e*) the models of their inventions, (*f*) the bibliography of their publications."

It will be inferred from what has been said in preceding chapters that the school cannot be classed among the wealthy institutions of the country. In its early days a considerable portion of the expense of its maintenance was borne by the founder; during the first eight years he expended more than $22,000 in it support. Upon the removal to the Van der Heyden mansion, in 1834, he built a laboratory and rooms for study upon the new site, and he continued to assist the institution until his death. Its equipment at first was not great, though it compared favorably with that used for scientific purposes in the oldest and wealthiest colleges. In 1828 the collections and library were valued at $3615 and the real estate at $1348. The total value of its property was $5009. The complete inventory made in 1846, after the removal to the Infant School lot, showed the total value of real estate, invested funds, library and apparatus to be $15,851 and the debts to amount to $1050. This value though small was not inconsiderable for schools of science at that period.

At various times the authorities of the school have

made appeals for aid to the Legislature of the State. One such petition, signed by B. Franklin Greene, LeGrand B. Cannon, John B. Tibbits and D. Thomas Vail, was presented shortly after the reorganization, and in the act making appropriations for general purposes, passed July 10, 1851, $3000 was given to the Institute. To aid in rebuilding after the fire, $10,000 was appropriated April 23, 1863. Another memorial signed by all of the trustees and by Director Charles Drowne was presented in 1866. They asked for $50,000. This was not given, but by an act passed April 23, 1864, the State Palæontologist was authorized to select from the duplicate fossils belonging to the State, and present to the Institute, a collection as full and complete as could be made. The fossils were given and an appropriation of $15,000 was also made May 8, 1868. Again in 1861, by an act passed April 28, $3750 was donated. These sums, together with the $744 received from the Regents between the years 1846 and 1853, while the Institute was under their visitation as an academy, make the total amount of money received from the State, since the foundation of the school, $32,494. This is wholly inconsiderable when compared with the sums which have been received from the same source by other institutions.

The aggregate amount of money subscribed at intervals since the reorganization by the trustees, alumni and citizens of Troy has been comparatively large. It would in general be invidious to mention

only those who have contributed most largely, though
in a few instances, where individual gifts have been
made for specific purposes, as in the case of the
Proudfit Observatory, Hart Professorship and Macdonald Prize, the names of the donors have been
given. It is gratifying to record the fact that in consequence of such contributions the property owned
by the school has been trebled during the past fifteen
years.

In the seventy years which have elapsed since the
foundation of the Institute, from 1824 to 1894, in-
clusive, there have been 1126 graduates. Of these
two hundred and sixty-five are known to be dead,
which would show that there are 861 living. This
number is probably somewhat too large, as there are
doubtless some dead, especially in the early years,
who have not been so recorded. Of these graduates
sixty-seven received the degree A.B. (r.s.), seventy-
seven that of Bachelor of Natural Science, B.N.S., nine
hundred and eighty-two graduated as Civil Engineers,
C.E., twenty-three as Mining Engineers, M.E., five
as Topographical Engineers, T.E., and the degree of
Bachelor of Science, B.S., was conferred upon thirty-
three. Eleven hundred and eighty-seven degrees
have therefore been conferred upon graduates.
Sixty-one of them took two degrees each. Fifty-
four of those who took two degrees were graduated
before the reorganization of 1849–50, and obtained
both C.E. and B.N.S.

Only four honorary degrees have been conferred :

In 1882 the honorary degree of Civil Engineer was conferred upon Charles H. Fisher, Chief Engineer of the New York Central and Hudson River Railroad, who had been a student in the class of 1853, and also upon Luiz da R. Dias, Chief Engineer of the Bahia and Caribaen Railroad, Brazil, who had been graduated in the class of 1860 as a Topographical Engineer. In 1884 the degree of Civil Engineer was also conferred upon William B. Cogswell, formerly of the class of 1852, the Chief Engineer and General Manager of the Solvay Process Company of Syracuse, N. Y. At the same time the honorary degree of Doctor of Philosophy was given James C. Booth, Director of the United States Mint at Philadelphia, who in 1831 had been a student at the school and an assistant to the Senior Professor.

The total number of students who have attended the Institute is not exactly known, though it closely approximates three thousand.

The reputation of the Institute as a school of engineering is well known. As the pioneer in any English-speaking country in this branch of education its fame was early established. Students have come to it from forty-two of the states and territories of the Union and from many foreign countries, including the Bahamas, Brazil, Canada, Chili, China, Costa Rica, Cuba, Ecuador, England, Germany, Honduras, Ireland, Italy, Japan, Mexico, Nicaragua, New Brunswick, Nova Scotia, Peru, Porto Rico, Russia, San

Domingo, Sandwich Islands, Spain, United States of Colombia and Venezuela.

Its renown has not been due to its age but to its methods of instruction, its rigid requirements for graduation and the work of its alumni.

Its methods and curriculums of the past have already been set forth in preceding chapters; those of to-day will follow.

Its requirements for graduation may be indicated in a general way by finding the ratio of the graduates in any class to the total number of students who have been members of it. Such ratios for every decade since the reorganization, beginning with 1860, are as follows : for the class of 1860 the percentage is 45.0, for 1870 it is 31.6, for 1880 it becomes 33.3 and for 1890 it is 27.0. The highest ratio, 50.0 per cent, is found for the class of 1885. In the class of 1874 it is 17.5 per cent. The average ratio for the last forty years is 36.0 per cent.

It is, however, to the work of its graduates that the reputation of the school is largely due. They have left an imprint in the history of the scientific development, constructive art and material progress of this and other countries which cannot be effaced. Their success has been marked not only in the profession of engineering and as scientific investigators but in business pursuits. It has been widespread. An appendix containing all the known addresses and occupations of the alumni was published, for the first time,

MEASURING THE VELOCITY OF A STREAM. 1894.

in the Register of November, 1860. This began
with the class of 1850. It has since been published
in all Registers and now includes all classes from
the beginning. A "geographical index", giving the
place of residence of all the graduates, was added in
1891. In the Register of 1894 this shows the living
ones to be at present at work in forty-five of the
states and territories of the Union and in nineteen
foreign countries.

In 1892 a pamphlet entitled " A Partial Record of
the Work of Graduates of the Rensselaer Polytechnic
Institute" was compiled from the annual Registers.
Its gives the names and positions of those of the
alumni whose pursuits could be easily classified.
Even in this particular it is necessarily incomplete,
and no mention is made in it of many who have
attained eminence in various diverse callings. This
partial list contains the names of thirty-three presidents, one hundred and twenty-one vice-presidents,
managers and superintendents, and sixty-nine chief
engineers of railroad companies, steel and iron works,
bridge companies, water works, electric companies,
mining companies, sewerage systems, canals, etc.
It shows that they have helped to build and operate
more than one hundred and nine thousand (109,000)
miles of the railroad system of North America alone
and that they have been connected as designers and
constructors with all the important bridge companies
and nearly all the great bridges of the country.

There is given also a list of fifty-six who have become professors in our leading universities, colleges and schools of science. More than two hundred have been connected with the American Society of Civil Engineers in its various grades of membership.

CHAPTER IX.

CURRICULUMS OF THE TWO EXISTING COURSES.

THE Institute has at present six buildings in use for purposes of instruction: the Main Building, the Winslow Laboratory, the Ranken House, the Williams Proudfit Astronomical Observatory, the Gymnasium and the Alumni Building. The Main Building contains lecture and recitation rooms, drawing-rooms and the laboratories of the department of Physics. The main hall of the institution, where the reading of theses and other general exercises takes place, is also in this building. The Winslow Laboratory is devoted entirely to the department of Chemistry. The first story contains rooms for quantative analysis and special investigations, and also the furnaces for assaying. The general laboratory for qualitative analysis and rooms for chemical balances and for the instructor in charge are on the second floor. The third story contains the general lecture-hall, a recitation-room, a room for the apparatus used in the lectures on general chemistry and an office for the use of the instructors in the department. In this room there is a carefully selected special chemical library.

The Williams Proudfit Observatory is well equipped with instruments for use in engineering instruction, containing a transit instrument, chronometer, chronograph, clocks and sextant. There is a special astronomical library in the computing-room. The Ranken House contains the machines used for testing wood, stone, cement and metals, and also a recitation-room for the department of mechanics. The gymnasium and Alumni Building have already been sufficiently described.

The methods of instruction are similar to those in vogue shortly after the reorganization. Text-books are largely used, though these are almost invariably supplemented by lectures. The classes are divided into small sections and each student is required to recite every day in all the subjects taught. Sometimes the recitations consist of interrogation only, but generally both interrogation and blackboard work are required every day.

The scholastic year is divided into two terms—the first beginning about the middle of September and the second about the first of February. For divisions D and A these are each about nineteen weeks in duration, and for divisions C and B the first term is about nineteen weeks and the second two weeks longer. This is because the latter two divisions spend the month of June in the field, surveying, whereas the former end the work of the second term about the middle of June. Each term is divided into three periods, the advance, the review and the examination. The advance, during which the student takes up a subject for the first time, lasts about fifteen weeks, the review about three, and the examination period is about one week in duration. In the review no new subject is studied, but those taken during the advance are repeated. During both the advance and review, when a subject is once taken up it is continued until it is finished. Recitations are held on consecutive days until the course is ended.

The principal course of instruction given is that of Civil Engineering, for which the degree conferred is Civil Engineer (C.E.). The instruction, however, is not narrowed to any special branch of civil engineering. The design of steam-engines, as well as that of bridges, sewerage systems, water-works,, etc. is taught and the student receives instruction as well in the principles of electrical engineering as in the location and construction of roads and railroads. A course in Natural Science is also given, upon the satisfactory completion of which the degree of Bachelor of Science (B.S.), is conferred. Special courses are given in Chemistry and special students are taken in any department for which they are qualified.

General schedules for the two courses will first be outlined, and afterwards detailed descriptions of the methods pursued in each department will be given.

SCHEDULE OF THE COURSE IN CIVIL ENGINEERING.

FIRST YEAR.

FIRST TERM.	SECOND TERM.
Solid Geometry.	Trigonometry.
Algebra.	Physics.
French.	French.
Projections, Theory.	Surveying, Theory.
Projections, Drawing.	Surveying, Practice.
Free-hand Drawing.	Colored Topography.
Plane Problems.	Bridge Drawing.
Elements of Drawing.	
Pen Topography.	

A Thesis must be written during the Summer vacation.

SECOND YEAR.

FIRST TERM.	SECOND TERM.
Physics.	Chemistry, Theory.
Logic.	Chemistry, Lectures.
Descriptive Geometry, Theory.	Differential Calculus.
Descriptive Geometry, Drawing.	Surveying, Theory.
Analytical Geometry.	Shades and Shadows, Theory.
Surveying, Theory.	Shades and Shadows, Drawing.
Surveying, Practice.	Perspective, Theory.
Physical Experiments.	Perspective, Drawing.
	Free-hand Drawing, Lettering.

A Thesis must be written during the Summer vacation. *A four-weeks' course in Surveying during the month of June is required.*

THIRD YEAR.

FIRST TERM	SECOND TERM
Integral Calculus.	Rational Mechanics.
Rational Mechanics.	Structures,
Geodesy.	Railroad Engineering, Theory.
Highway Engineering.	Astronomy.
Chemistry, Qualitative Analysis.	Machine Construction, Theory. ·
Mineralogy.	Machine Construction, Plates.
Electricity and Magnetism.	Chemistry; Blowpipe Analysis;.
Map Drawing.	Assaying.

A Thesis must be written during the Summer vacation. *A four-weeks' course in Railroad Engineering during the month of June is required.*

ΓOURTH YEAR.

FIRST TERM.	SECOND TERM.
Machines.	Bridge Design.
Resistance of Materials	Hydraulics.
Hydraulics.	Hydraulic Motors.
Sewerage.	Thermodynamics.
Bridges and Roofs.	Steam Engineering.
Economic Theory of Railroad Location.	Stone Cutting, Theory.
	Stone Cutting, Plates.
Practical Astronomy, Theory.	Electrical Engineering.
Practical Astronomy, Observations.	Physical Laboratory Work.
	Geology.
Metallurgy.	Law of Contracts.
Physical Laboratory Work.	

A Graduating Thesis must be presented.

SCHEDULE OF THE COURSE IN NATURAL SCIENCE.

The studies of the course in Natural Science are identical with those in Civil Engineering during the first two years.

THIRD YEAR.

FIRST TERM.	SECOND TERM.
Calculus.	Astronomy.
Electricity and Magnetism.	Geology, Lithology.
Mineralogy, Petrography.	Histology.
Map Drawing.	Chemistry, Organic ; Blowpipe Analysis, Assaying.
Chemistry—Qualitative Analysis ; Elementary Quantitative Analysis.	

A Thesis must be written during the Summer vacation

FOURTH YEAR.

FIRST TERM.	SECOND TERM.
Metallurgy—General Metallurgy, Iron Metallurgy.	Physical Laboratory Work.
	Paleontology.
Chemistry—Quantitative Analysis ; Analysis of Commercial and Industrial Products.	Mineralogy, Determinative.
	Petrography.
Physical Laboratory Work.	Chemistry — Quantitative Analysis; Volumetric and Gravimetic Analysis.
	Law of Contracts.

A Graduating Thesis must be presented.

Mathematics and Astronomy.—The subjects in this department are taught partly by interrogation and explanation and partly by exercises at the black-board. The classes are divided into small sections so that each student recites every day and receives the constant personal attention of the instructor. A careful record is kept of his daily work. During the first year thorough instruction is given in solid geometry, higher algebra and trigonometry. These are followed by analytical geometry and differential calculus in the second year, and by integral calculus in the third. Lectures on the theory and various forms of the slide-rule are also delivered. In all these subjects examples of a practical nature are constantly given. The text-books used are supplemented by notes prepared by the instructors.   '

The course in descriptive astronomy is given in the third year, and that in spherical and practical astronomy in the fourth. In the latter are considered the adjustment and use of portable instruments, correction of observations, determination of time, latitude, longitude and the meridian, the methods of least squares and similar subjects. The theory is supplemented by work in the observatory, where the use of the sextant, chronograph, transit instrument, etc., is taught.

Descriptive Geometry and Stereotomy.—In this department careful and thorough instruction is given in free-hand drawing, lettering, the use of drawing instruments, tinting, shading, isometric and ortho-graphic projections, tracing and making blue-prints,

MERIDIAN OBSERVATIONS IN THE OBSERVATORY.   1894.

the theory and practice of shades, shadows and perspective, machine construction and drawing, including gearing and the slide-valve, and stone cutting. In all these subjects a great amount of time is spent in the drawing-room under the immediate supervision of the instructor, and original work sufficient to fix the principles is required. In descriptive geometry, for instance, although a lesson is assigned for each day from the text-book the student is seldom given a problem found there, but is required to prove an original one illustrating the same principles. Besides the drawing required in the course in stone cutting, plaster of paris models of arches, stairways, etc., are constructed by the students.

The drawing-rooms are commodious and well equipped with all that is necessary for the student. A large number of models of joints in wood, the projections and intersections of solids ; groined, cloistered and other arches, by Schröder, and others of the different solids and warped surfaces, as cylinders, cones, conoids, hyperboloids, hyperbolic paraboloids, etc., with their intersections and tangents, after those of Olivier, are in possession of the department. These, with valves, machines and parts of machines furnish sufficient material for illustration and use.

Chemistry.—The course in chemistry, which is obligatory for all students, consists of daily lectures, during the last part of the second year, upon general inorganic chemistry. These are accompanied by daily recitations, including the solution of chemical problems.   Each student must recite every day.   In preparing for this recitation in general chemistry the student is expected to make use of the chemical cabinet, in order that he may be familiar with the appearance and some of the simpler properties of the materials referred to in the class-room.   From time to time students are called upon to prepare and deliver short lectures upon assigned subjects of technical interest.

The course in qualitative analysis extends over the first half of the third year, with laboratory work five days in each week.   During this course the student acquires ability to analytically examine all the ordinary materials likely to be presented to his attention during his professional engineering practice. He is, as far as possible, given charge of outside questions which come to the laboratory for solution. Blowpipe analysis and assaying extend over part of the second term of the third year, particular attention being given to the assay of gold and silver and to the recognition of such ores of the heavy metals as may be met with in the mining regions of this country.

Quantitative analysis and organic chemistry are not given to candidates for the degree of Civil Engineer.   Courses in these subjects are given to candidates for the degree of Bachelor of Science, to post-graduates and to special students.   Very complete arrangements make these courses especially thorough.   The examination of water for public and domestic supply, for boiler purposes and for use in manufactures is made a specialty in this department.

The laboratory is well equipped with material and apparatus for the course in general and analytical chemistry outlined above. Fine analytical and assay balances, filtering apparatus, large and small gas-ometers, microscopes, spectroscopes, thermometers, areometers, batteries, etc., are in constant use. The fire-rooms for assaying are provided with pot-fur-naces for the crucible assay of the base metals and with muffles for the assay of gold and silver. Abun-dant specimens for illustrating the lectures on tech-nical chemistry are also to be found, the chemical cabinet containing a valuable collection of glass, earthenware, porcelain, gunpowder, paper, coal-tar products, etc.

Mineralogy, Geology and Metallurgy. — These subjects are given by text-books, lectures and inter-rogations. They are well illustrated by the very large collection of specimens, maps and charts in the possession of the Institution. The metallurgical col-lection contains a very large number of ores and their products. Steel and iron works in the neigh-borhood are inspected during the period of instruction.

Physics.—The work of this department begins in the last term of the first year with the mechanics of solids, liquids and gases, and acoustics. Optics and heat are studied during the first term of the second year, and electricity and magnetism during the first term of the third year. These subjects are developed by daily lectures. The student uses a text-book, and is held strictly accountable for an exact knowledge of its contents, but much instruction is given additionally in the lectures, accompanied with full experimental illustrations. He is required to take notes during the course of the lectures and to copy others which have been put upon the blackboards. In the course of daily recitations problems are frequently assigned, and upon these, as well as on demonstrations of theory, the student is required to give both oral and written explanations. During the first term of the second year a course of laboratoy work is conducted in which the student is introduced to the methods of quantitative measurement, and he thus acquires some familiarity with the use of physical instruments. For each exercise due preparation is made by appropriate reading and a report is written which is examined by the instructor. During the first and second terms of the fourth year laboratory practice is continued, prominence being given to methods in electrical and magnetic measurement.

During the second term of the fourth year a course in thermodynamics is given, and this is followed by lectures on the elements of electrical engineering as an accompaniment to the laboratory work in electrical measurement.

The equipment of the lecture-room and laboratory includes such apparatus as Kœnig's tuning-forks, siren, organ-pipes, vibratory plates, rods, sonometer ; large air-pump with accessories ; Atwood's machine ;

MEASURING THE RESISTANCE OF INCANDESCENT LAMPS. 1894.

compound microscope with accessories ; cathetom-
eter ; Bunsen photometer ; Morton projecting lantern with accessories, including four hundred slides
illustrating various subjects in physics, etc. ; diffrac-
tion gratings ; prisms, lenses, etc. ; large Hilger
spectrometer ; Nicol's prism, two-and-a-half-inch aperture ; electric batteries of various types ; thermo-
pile ; Thomson's galvanometer, four thousand ohms ;
D'Arsonval galvanometer ; condenser, one-third mi-
crofarad ; Mascart electrometer ; Weston voltme-
ter, Weston ammeter and milliammeter ; large in-
duction coil ; Ruhmkorff electromagnet ; Toepler
electric influence machine ; Carhart-Clark standard
cell ; Elliott resistance coils and Wheatstone bridge ;
etc. A gas-engine and dynamo supply the electricity
required, and the laboratory is also provided with a
battery of storage-cells.

Surveying.—The student begins the work in surveying during his first year at the Institute. In the
second term of this year he is taught the use of the
chain, tape and compass. He also makes a compass
survey of a small piece of land which is mapped and
the area computed. In the second year the construction and use of all modern surveying instruments,
including transit, level, solar compass and attachment,
clinometer, hand-level, aneroid barometer, planimeter,
etc., are taught in the class-room, as are also the various methods in modern use of making land, topographical, hydrographical, mine and city surveys. In
topographical surveying, while all methods are taught

and the conditions rendering one method more suitable than another, particular attention is paid to the transit and stadia, and the students become thoroughly familiar with this most generally suitable method. During the first term daily practice in the adjustment and use of the various instruments before enumerated is given. Surveys of limited extent are executed, a meridian is established with the solar compass, checked by stellar observations, and the magnetic declination of the needle determined.

At the close of the year the class is taken into the field for four weeks, and makes a complete topographical survey of an area selected with reference to the diversity of problems it presents. This survey is also made to include hydrographic work, as the portion of the stream within the area chosen for work is mapped from soundings and its flow determined. For this work the Institute has a large equipment, including compasses, transits, levels, solar compass, plane-table, rods, hand-levels, clinometers, barometers, etc.

Geodesy.—Besides the course in astronomy, in which the students are taught to determine latitude, longitude, time, etc., from observations on the heavenly bodies, a brief course in geodetic surveying is given in the third year. The work includes the methods of measuring base lines, field-work of triangulation, adjustment of triangles and quadrilaterals and a discussion of the figure of the earth.

Highway Engineering.—During the third year

there is given a course in highway engineering, in which is discussed the location, construction and maintenance of country roads and city streets, the advantages and disadvantages of the various paving materials and specifications for each, and a study is made of the various road laws in force and their adequacy.

Railroad Engineering.—The subject of railroad engineering is begun in the third year with a theoretical course in railroad curves, turnouts and minor structures, and the staking out and computation of railway earthwork. The course also includes a discussion of the method of making railway location surveys, and a contour map is furnished the student on which he projects a location line and makes an estimate of materials and cost. This theoretical course is followed at the close of the year by four weeks of field practice in railroad surveying, during which a preliminary survey is made and mapped, a location projected and run in, the work staked out, quantities computed and cost estimated. A line from three to eight miles in length is usually located, and through the courtesy of railroad officials interested in the Institute the classes not infrequently have an actually contemplated line to examine, which secures an interest and faithfulness not always obtained on a mere " practice " line.

In the fourth year the subject generally know as Economic Theory of Railroad Location, embracing the items of train resistance and the effect of grades, curves and length of line on operation, is thoroughly studied, together with the correlative subjects of economic construction, maintenance of way, etc. Numerous problems are given to illustrate the subject, and a short thesis comparing two or more possible locations for a line, the data for which are given, is written. The students also discuss in the light of the new knowledge the location made the previous year. In addition to the above there is given in the fourth year a comprehensive series of lectures on railway signals, embracing the construction and operation of block signals and interlocking signals for yards, crossings, etc.

Summer Courses.—The summer courses in surveying in the second and third years are particularly valuable, on account of the continuous and practical character of the work. The student is employed all day for six days in the week, and the methods used both in the topographical and railroad surveys embody the latest modern practice. The work is usually located in the Adirondack foot-hills, and forms the most enjoyable and heathful, as well as valuable, portion of the surveying instruction. These courses are open to a limited number of special students who show themselves competent to perform the work.

Topographical Drawing.—This subject is taught in the first, second and third years, of the course. In the first year the student learns to make the various topographic symbols, both in pen and ink and in color. In the second year, in connection with the course in surveying, he maps small areas from notes furnished him, measures and computes the areas and draws contours, projects grades and computes volumes of earthwork involved in surface grading. He also makes the skeleton map of the summer survey. In the third year he completes this map and also makes, in the field, the map of the railroad survey. The use of the planimeter and the various diagrams for estimating areas and earthwork are taught.

Rational Mechanics.—At the conclusion of the course in integral calculus during the first term of the third year instruction in rational mechanics begins. In this course, which extends over a part of two terms, with recitations and lectures every day, the fundamental theoretic principles of statics, cinematics and dynamics, which underlie and form the foundation of all branches of engineering, are taught. These include the resolution and composition of forces, the determination of the centre of gravity and moment of inertia of various bodies and cross-sections in constant use in practical work, the principles of internal stress, translation and rotation, momentum, impulse, energy, work, impact, oscillation, fluid pressure, etc. The higher treatises and text-books, supplemented by notes, are used. The method of instruction, which applies as well to the technical subjects in the department of mechanics as to the rational, is as follows : The class is divided into sections and each section, after a combined lecture and thorough interrogation by the Professor in charge, goes to the Assistant for a recitation on certain selected parts of the subject. The Assistant requires each student each day to put one of these articles on the blackboard and explain it. During this explanation he is interrogated upon the principles involved.

Structures.—The theory of structures is taught during the last term of the third year. This includes the equilibrum and stability of frames, cords, arches, buttresses, chimneys, abutments, piers, retaining-walls, dams, etc. Analytical and graphical methods of treatment are elaborated. A treatise on masonry construction is also used as a text-book, and the strength, properties and cost of cement, mortar, concrete, brick and stone masonry, together with all the more important kinds of foundations, are considered.

Resistance of Materials.—The elasticity and resistance of the materials of engineering are considered during the first term of the fourth year. The fundamental equations of the theory of flexure are first determined and applied to a consideration of the strength of simply supported and continuous beams and of columns. Practical formulæ for the strength of beams are determined and the right-line long-column formula, and those of Gordon and Euler, are deduced. Attention is also paid to the graphical representation of the strength of columns. Theoretical formulæ for torsion are developed and applied to a consideration of the strength of shafting. The design of riveted joints for boiler and tube work is

TESTING METALS.   1894.

taken up and the proper size and pitch of rivets determined.

In the practical part of the subject the coefficient of elasticity, elastic limit, ultimate resistance and other properties of cast and wrought iron, malleableized iron, steel, bronze, copper and other metals in tension, compression and shear are studied, and the students are required to make experiments on the testing-machine and determine their properties as above outlined. The value of wood, stone, brick, etc., for use as materials of engineering is investigated, and each student also determines the strength of cement by the use of a cement-testing machine. Attention is paid to the fracture and appearance of metals and also to the effect of repetition and reversal of stress.

Bridges and Roofs.—The course on bridges and roofs is given in the first and second terms of the fourth year. The first part is devoted to the theory of stresses. In this the student becomes familiar with the calculation of stresses in plate girders, in Howe, Pratt, Whipple and lattice bridges and in trusses with curved chords; also in cantilever, suspension and drawbridges, and in various kinds of roof trusses. Analytical and graphical methods and the method of wheel concentrations and of panel loads are used. Details and connections are carefully considered and studied from the very large collection of blue prints of existing structures of all kinds in possession of the Institute. A set of bridge specifications forms a part of the course, upon which recitations are required, and hand-books of bridge and iron works are used for reference. During this course the class is taken out for an examination and comparison of various styles of bridges in the vicinity, and a bridge shop is also visited and the machines and methods of manufacture explained.

The second part of the course in the second term is taken up with the design of bridges and parts of bridges. The student makes all the calculations and complete shop drawings of the work in hand, each bridge being different from the others, and tracings and blue prints are finally made.

Hydraulics.—This subject is taught in the fourth year. It includes hydrology, hydrostatics, theoretical hydraulics, the flow of water through orifices, over weirs and dams, through tubes and pipes, and in conduits, canals and rivers, the measurement and cost of water-power, the dynamic pressure of flowing water, hydraulic motors and the general principles of naval hydromechanics. Numerous examples illustrating the principles are given. In the direction of water-supply engineering there are considered general rainfall statistics, precipitation, evaporation, the collection and storage of water and its impurities; the practical construction of water-works, including reservoir embankments, waste-weirs, partition-walls, conduits, distributing systems and the various methods of filtering. The delivery of water by pumps is here touched upon, though this matter is more thoroughly treated in the course on the steam-engine. The theory and efficiency of the various forms of water-wheels are investigated and the students are instructed with regard to the different kinds of turbines, with their draught-tubes, diffusers and governors.

They are required to measure the flow of adjacent streams by means of weirs, and thus practically to find the discharge. Practice in the measurement of the velocity of streams by means of current meters and floats is also given, and models of valves, motors, practical working turbines, etc., add value to the instruction.

The subject of aerodynamics is also taken up in this course and the flow of air through orifices and in pipes, blowing-engines, the relations between the velocity and pressure of the wind, anemometers, windmills, etc., are studied.

Sewerage Systems.—The design of sewerage systems is taken up in the fourth year. A comparison of the cost and efficiency of the different systems is made and the conditions under which each should be used explained. The various methods of sewage disposal are exemplified and their efficiency discussed. The effect of the surface slope and magnitude of area drained in connection with the maximum rainfall is considered and main and branch sewers for the separate and combined systems are proportioned and their cost determined. The materials of construction, foundations required, methods of laying and descriptions of details, such as branches, man-holes, catch-basins, etc., are also given.

Steam Engineering.—The course in steam engineering is given during the last term of the fourth year. It consists of a series of lectures by a well-known consulting mechanical engineer. The properties of steam are first elaborated, and afterwards the details and construction of the various engines and boilers in ordinary use considered. The strength of their parts are calculated and their general operation explained. The course also includes pumping machinery. The lectures are illustrated by drawings, photographs and hand-books, and books of reference are used for consultation. Each student makes a general design for a locomotive, pumping, marine or other form of engine, though detailed drawings are not expected. He is also required to take indicator diagrams from some engine and determine from them its power. Examinations of various forms of steam-engines in the vicinity are also made under the direction of the instructor.

Theses.—A thesis on some technical subject must be written by each student during each summer vacation.

A graduating thesis, which must be either a review of or a design for a machine, structure, plant, system, or process belonging to a department of scientific or practical technics, is also required. In the department of Civil Engineering designs are generally required, while in that of Natural Science special investigations of a scientific nature are expected. The titles of some recent theses follow :

Design for a Railroad Bridge with Curved Upper Chord and Secondary System. Span 400 feet.

Design for a Wrought Iron Railroad Viaduct with Spans of 60 and 30 feet.

Design for a Through Pratt Truss Railroad Bridge. Span 175 feet.

Design for a Bowstring Highway Bridge with a Buckle-plate Floor. Span 160 feet.

Design for the Roof of a Factory. Span of Trusses 120 feet.

Design for the Steel Skeleton of a Building 75 feet by 125 feet, 11 Stories High.

Design for an Additional Water-supply for the City of Troy. Gravity system.

Design for the Distributing System and Stand-pipe for a Town of 5000 Inhabitants.

Design for a Driven-well System of Water-works for a City of 20,000 Inhabitants.

Design for a Turbine Plant developing 2500 Horse-power under a Head of 45 feet.

Design for a System of Sewers for a Town of 50,000 Inhabitants.

Design for a Stone Arch Railroad Bridge. Two Spans of 60 feet each.

Design for the Substructure of a Railroad Bridge, with different conditions, viz : On Clay ; using Coffer Dam ; Piles ; Caisson.

Design for a Steel Dam 100 feet High, with a discussion of the Relative Cost of Steel and Masonry Dams.

Design for a Pair of Triple-expansion Marine Engines to develop 10,000 Indicated Horse-power.

Design for the Blooming-mill of a Steel Rail Plant, capacity 1000 Tons per diem.

Design for a Hydraulic Plant to develop 400 Horse-power under a Head of 250 feet and the Electric Transmission of this Power for a Distance of four Miles.

Design for a Dynamo to Supply Power for Operating a Street Railway.

Design for a Double-U Magnet Motor of 15 Horse-power.

Design for an Interlocking Switch and Signal System for the D. and H. Yards at Green Island, N. Y.

The Hardening of Steel and its Effect upon the Ultimate Resistance.

The Effect of Overstrain on Metals.

Research on the Oxidation of Organic Matter in Potable Water.

The Manufacture of Super-phosphates.

These titles illustrate the general character of the course and the capacity of students who have taken it to deal with diverse problems of an engineering and scientific nature. Theses for which the Macdonald prize has been awarded have been criticised in a most favorable manner by some of the most eminent American engineers.

WEIR MEASUREMENTS. 1894.

BIBLIOGRAPHY.

ALL the publications and records, known to the writer, relating to the school from the time of its foundation to the present date are given in the following list. More or less use has been made of most of them in collecting the information given in this history.

Minutes of the Board of Trustees, 1824–1849, and 1862–1894.

Laws of the State of New York: 1826, Chapter 83 ; 1832, Chapter 327 ; 1835, Chapter 254 ; 1837, Chapter 351 ; 1850, Chapter 49 ; 1851, Chapter 498 ; 1861, Chapter 151; 1863, Chapter 210; 1864, Chapter 320; 1866, Chapter 229; 1868, Chapter 717; 1871, Chapter 869, and 1887, Chapter 277.

Preparation Branch recently established at Rensselaer School. September 14, 1826.

This consists of a single sheet, and is published in full in Chapter IV.

Constitution and Laws of Rensselaer School, in Troy, New York ; adopted by the Board of Trustees April 3, 1826 ; together with a Catalogue of Officers and Students. Albany, 1826. 8vo, 28 pp.

Rensselaer School Exercises in the Fall, Winter and Spring Terms, including those of the Preparation and District Branches. Published under the Direction and Authority of the Board of Trustees by the Senior Professor. June 27, 1827.

This forms pages 29 to 48 of the pamphlet dated April 3, 1826, above mentioned.

Triennial Catalogue of the Officers and Members of Rensselaer School, Troy, N. Y., 1828. 8vo, 15 pp.

To Graduates of Colleges and Teachers of Academies and of Common Schools. October 29, 1828. 4to, 1 p.

Rensselaer School extended. September 23, 1829. 4to, 1 p.

Rensselaer School Flotilla for the Summer of 1830. January 28, 1830. 8vo, ⌐ pp.

Exercises of Rensselaer School, with an Account of its Origin and Characteristics. Also a Catalogue of Officers and Students, 1831. 8vo, 24 pp.

Rensselaer School Notices for the Eighth Annual Course of Instruction, 1831 and 1832. 8vo, 4 pp.

Rensselaer Institute, Troy, N. Y., Notices for the Ninth Annual Course, 1832 and 1833. 8vo, 4 pp.

A Digest of the Laws and Rules of Exercise and Discipline in Rensselaer Institute. With a Triennial Catalogue. 1833. 8vo, 40 pp.

Synopsis of the Mathematical Course of Instruction at Rensselaer Institute from November 19, 1834, to February 11, 1835. 4to, 1 p.

Notices of Rensselaer Institute, Troy, N. Y. October 14, 1835. 8vo, 4 pp.

This is printed in full in Chapter VI.

Periodical Notices of Rensselaer Institute. To Engineers, Geologists, Chemists, Naturalists, etc. 1838 and 1839. 8vo, 8 pp.

Rensselaer Institute, 35th Semi-annual Session for 1841–2. 8vo, 8 pp.

Rensselaer Institute, 37th Semi-annual Session for 1842–3. 8vo, 8 pp.

Catalogues for the years 1844, 1845, 1846, 1847, and 1849.

Announcement of the Fifty-second Semi-annual Session of the Polytechnic Institution at the City of Troy. April 15, 1850. 4to, 4 pp.

Programme, etc., of the Rensselaer Institute : a Polytechnic Institution at the City of Troy, N. Y. February 1, 1851. 8vo, 8 pp.

To the Legislature of the State of New York. 3vo, 8 pp. This is a memorial signed by a committee composed of B. Franklin Greene, Le Grand B. Cannon, John B. Tibbits and D. Thomas Vail.

Annual Registers of the Institute, dated as follows : August, 1851; October, 1852; August, 1854; August, 1855 ; March, 1857; May, 1858 ; 1859 ; August, 1859; 1860; November, 1860 ; July, 1861; July, 1862; November, 1862; July, 1863; 1863–64, First Term ; 1863–64, Second Term ; July, 1864 ; 1864–65, First Term; July, 1865; November, 1865; July, 1866 ; October, 1866; July, 1867; December, 1867; July, 1868; November, 1868; July, 1869; December, 1869; April, 1870; July, 1870; July, 1871; June, 1872; September, 1872; December, 1872; July, 1873; December, 1873; May, 1874; July, 1874; July, 1875; November, 1875; August, 1876; February, 1877; July, 1877; 1877–78; July, 1878; April, 1879; November, 1879 ; June, 1880; November, 1880; July, 1881; January, 1882; November, 1882; February, 1883; October, 1883; April, 1884; February, 1885; July, 1885; June, 1886; July, 1886; January, 1887; November, 1887; May, 1888; May, 1889; November, 1889; June, 1890; 1891; March, 1892; April, 1893; March, 1894.

Report of the Committee appointed by a Public Meeting of the Citizens of Troy, on the Subject of certain Proposed Improvements of the Rensselaer Polytechnic Institute. January 21, 1854. 8vo, 4 pp. Signed by Thomas W. Blatchford, J. M. Warren and John A. Griswold.

The Rensselaer Polytechnic Institute : Its Reorganization in 1849–50; Its Condition at the Present Time; Its Plans and Hopes for the Future. By the Director of the Institute. Also, the Statement of a Committee, appointed by and of the Trustees of the Institute, for the Presentation of its Various Interests to the Citizens of Troy. May 10, 1856. 8vo, 87 pp.

The first part, written by B. Franklin Greene, includes 80 pages of the pamphlet; the statement of the committee and table of contents takes up the remainder. The committee was composed of Hiram Slocum, John A. Griswold, Joseph M. Warren, Jonathan Edwards, Thomas C. Brinsmade, John B. Tibbits, Jonathan E. Whipple and B. Franklin Greene.

The " Transit ", an annual issued under the auspices of the Fraternities. Volumes 1 to 29, extending from 1865 to 1894.

Memorial of the Rensselaer Polytechnic Institute, Troy, N. Y., to the Legislature of the State of New York, 1866. 8vo, 8 pp.

This was signed by all the members of the board of trustees and by Director Charles Drowne.

Report of a Committee of the Trustees of the Rensselaer Polytechnic Institute concerning the System of Instruction, with Proposed Modifications. 1870. 4to, 43 pp. E. Thompson Gale, Alexander L. Holley and Clarence E. Dutton, committee.

Papers relating to the Organization and to the First Regular Meeting of the Association of Graduates of the Rensselaer Polytechnic Institute, held at Troy, N. Y., June 22 and 23, 1869. 1870. 8vo, 24 pp.

History of the Winslow Laboratory. By Henry B. Nason. June 15, 1874. 8vo, 13 pp.

Proceedings of the Semi-centennial Celebration of the Rensselaer Polytechnic Institute, Troy, N. Y., held June 14–18, 1874, with Catalogue of Officers and Students, 1824–1874. 1875. 8vo, 223 pp.

This was edited by Henry B. Nason.

Meeting of Alumni in New York, February 18, 1881. Proceedings of the First Reunion of Graduates. 1881. 8vo, 30 pp.

The " Polytechnic ", a publication issued by the students of the Institute each month during the scholastic year. February, 1885, to date (1894).

Biographical Record of the Officers and Graduates of the Rensselaer Polytechnic Institute, 1824–1886. By Henry B. Nason. 1887. 8vo, 614 pp.

Distant Examinations for Admission. Report of Alumni

Committee and Petition to Board of Trustees. January, 1888. 8vo, 23 pp. Warren T. Kellogg, De Volson Wood, S. W. Barker, Richard P. Rothwell, P. H. Baermann, A. P. Boller and I. A. Stearns, committee.

Central R. P. I. Association. Constitution and By-laws. Edgar B. Kay, Secretary, February, 1888. 8vo, 8 pp.

Rensselaer Society of Engineers. List of Members, 1892. 8vo, 23 pp.

A Partial Record of the Work of Graduates of the Rensselaer Polytechnic Institute, Troy, N. Y. 1892. 8vo, 27 pp.

The Rensselaer Polytechnic Institute, Troy, N. Y., Founded 1824. Handbook of Information. 1893. 8vo, 23 pp.

APPENDIX.

TRUSTEES, INSTRUCTORS, AND GRADUATES

OF THE

RENSSELAER POLYTECHNIC INSTITUTE

FROM

1824 to 1894.

* Indicates those known to be deceased.

Patron.

* Hon. Stephen Van Rensselaer........................... 1824–39

Presidents.

* Rev. Samuel Blatchford, D.D........................... 1824–28
* Rev. John Chester, D.D................................ 1828–29
* Rev. Eliphalet Nott, D.D., LL.D 1829–45
* Rev. Nathan S. S. Beman, D.D., LL.D.................. 1845–65
* Hon. John F. Winslow................................. 1865–68
* Thomas C. Brinsmade, M.D............................. 1868–68
* Hon. James Forsyth, LL.D............................. 1868–86
   Hon. John Hudson Peck, LL.D.......................... 1888

Vice-Presidents.

* Orville L. Holley, First Vice-President.................. 1824–41
* T. Romeyn Beck, M.D., Second Vice-President 1824–29
* Hon. David Buel, Jr., Second Vice-President............ 1829–60
* Rev. Nathan S. S. Beman, D.D., LL.D.................. 1841–45
* William P. Van Rensselaer............................. 1845–65
* Thomas C. Brinsmade, M.D. 1865–68
* Hon. George Gould.................................... 1868–68
* E. Thompson Gale, C.E............................... 1868–72

163

* Hon. William Gurley, C.E............................ 1872–87
  Albert E. Powers................................... 1887

Secretaries.

* Moses Hale, M.D.................................... 1824–37
* Rev. Mark Tucker, D.D.............................. 1837–38
* Rev. Erastus Hopkins............................... 1838–41
* Hon. Isaac McConihe, LL.D.......................... 1841–42
  Hon. Joseph White, LL.D............................ 1842–49
* Stephen Wickes, M.D................................ 1849–54
  Rev. John B. Tibbits, A.M.......................... 1854–61
* Hon. William Gurley, C.E........................... 1861–71
  William H. Doughty, C.E............................ 1871

Treasurers.

* Hon. Hanford N. Lockwood........................... 1824–44
* Thomas C. Brinsmade, M.D........................... 1844–47
* Hon. Day Otis Kellogg.............................. 1847–50
  William H. Young................................... 1850

Trustees.

* Rev. Samuel Blatchford, D.D........................ 1824–28
* Elias Parmelee, A.M................................ 1824–34
* Hon. John Cramer.................................. 1824–49
* Hon. Guert Van Schoonhoven......................... 1824–44
* Hon. Simeon De Witt................................ 1824–28
* T. Romeyn Beck, M.D., LL.D......................... 1824–28
* Hon. John D. Dickinson, LL.D....................... 1824–40
* Jedediah Tracy..................................... 1824–25
* Hon. Richard P. Hart............................... 1825–43
* Gen. Nicholas F. Beck, A.M......................... 1828–31
* Judge Jesse Buel................................... 1828–35
* Philip S. Van Rensselaer, A.M...................... 1833–43
* Rev. Phineas L. Whipple............................ 1833–37
* Hon. George Tibbits,   *ex officio* Mayor........ 1835–36
* William D. Haight,   "   "   Alderman..... 1835–36
* John P. Cushman,   "   "   Recorder..... 1835–38
* James Wallace,   "   "   Alderman..... 1836–38
* Hon. Jonas C. Heartt,   "   "   Mayor........ 1837–43
* Elias Dorlon,   "   "   Alderman..... 1838–39
* H. W. Strong,   "   "   Recorder..... 1838–44
* Henry Everts,   "   "   Alderman..... 1839–40
* Livy S. Stearns,   "   "   Alderman..... 1840–41
* Henry Everts   "   "   Alderman..... 1841–42

* Rev. W. B. Sprague, D.D............................. 1841–44
* John Holme 1841–56
* Rev. Alva T. Twing, D.D............................. 1841–67
* Hon. David Buel, Jr............................... 1842–44
* Rev. Eliphalet Nott, D.D., LL.D 1842–45
* Rev. Nathan S. S. Beman, D.D., LL.D 1842–65
* Hon. Isaac McConihe, LL.D.......................... 1842–67
* Daniel G. Egleston,   *ex officio* Alderman............ 1842–44
* Hon. Gurdon Corning,   "   "   Mayor 1843–47
* Abram B. Olin, LL.D.,   "   "   Recorder 1844–50
* Jared S. Weed,   "   "   Alderman............ 1844–45
* Rev. Reuben Smith................................ 1844–47
* Thomas C. Brinsmade, M.D........................... 1844–68
* William P. Van Rensselaer......................... 1845–65
* Luther Tucker................................... 1845–49
* Hon. Daniel D. Barnard, LL. D...................... 1845–48
* Stephen Bowman,   *ex officio* Alderman...... 1845–47
* James Dana,   "   "   Alderman...... 1847–49
* Hon. Francis N. Mann, A.M.,   "   "   Mayor......... 1847–50
* Stephen Wickes, M.D.............................. 1847–54
* W. T. Seymour................................... 1848–49
* Benjamin P. Johnson.............................. 1849–66
* Alexander Van Rensselaer, M.D..................... 1849–67
* John Wilkinson.................................. 1849–55
  Hon. Joseph M. Warren, A.M....................... 1849
  Le Grand B. Cannon.............................. 1849–64
* Hiram Slocum................................... 1849–60
* Orsamus Eaton.................................. 1849–59
  Rev. John B. Tibbits, A.M......................... 1849–67
* Leonard McChesney, *ex officio* Alderman.............. 1849–50
* Amos Dean, LL.D................................ 1849–53
* D. Thomas Vail, A.M.............................. 1850–82
  Hon. Joseph White, LL.D.......................... 1850–55
* Hon. Day Otis Kellogg,   *ex officio* Mayor.......... 1850–50
* Hon. Hanford N. Lockwood,   "   "   Mayor......... 1850–51
* Hon. George Gould.............................. 1852–53
* Hon. Foster Bosworth 1853–53
* Hon. Elias Plum................................ 1853–54
* Thomas W. Blatchford, M.D 1854–66
* Hon. Jonathan Edwards........................... 1854–67
* Hon. John A. Griswold, *ex officio* Mayor.............. 1855–56
  B. Franklin Greene, C.E., A.M...................... 1855–59
* Hon William Gurley, C.E.......................... 1855–87
* Hon. Jonathan E. Whipple........................ 1856–66
* Hon. Hiram Slocum, *ex officio* Mayor................. 1856–57

* Hon. Alfred Wotkyns, M.D., *ex officio* Mayor......... 1857–58
* Hon. Arba Read,   "   "   Mayor......... 1858–60
* Hon. John F. Winslow............................... 1860–68
* E. Thompson Gale, C.E.... 1860–87
* Hon. John A. Griswold..................... 1860–72
  Hon. Isaac McConihe, Jr. *ex officio* Mayor 1860–61
  Hon. George B. Warren, Jr., "   "   Mayor.......... 1861–62
  William H. Young 1861
* Hon. Lyman Wilder................................... 1861–85
* Hon. Arba Read..................................... 1861–63
  Albert E. Powers 1861
* Rev. Peter Bullions, D.D.............................. 1862–64
* Hon. James Thorn, M.D.,   *ex officio* Mayor........ 1862–63
* Hon. William L. Van Alstyne, "   "   Mayor........ 1863–64
* Hon. James Thorn, M.D.,   "   "   Mayor........ 1864–65
* Rev. Duncan Kennedy, D.D........................... 1864–67
* Hon. Jonas C. Heartt................................ 1864–74
* Hon. George Gould.................................. 1864–68
* David Cowee,...................................... 1865–87
* Alexander L. Holley, LL.D........................... 1865–66
* Hon. Uri Gilbert, *ex officio* Mayor..................... 1865–66
* Frederick B. Leonard, M.D........................... 1866–71
  James S. Knowlson, A.M.............................. 1866
* Hon. Uri Gilbert................................... 1866–88
  Hon. David A. Wells, LL.D., D.C.L................... 1866–76
* Hon. John L. Flagg, *ex officio* Mayor.................. 1866–68
  Hon. Charles R. Ingalls.............................. 1868
  Rev. Marvin R. Vincent, D.D......................... 1867–69
  William A. Shepard.................................. 1867–83
* Hon. James Forsyth, LL.D........................... 1867–86
* Joseph W. Fuller,.................................. 1867–89
  Hon. William Kemp................................. 1867
* Hon. Francis S. Thayer............................... 1868–80
* Azro B. Morgan 1868–71
  Hon. Miles Beach, *ex officio* Mayor.................... 1868–70
  Rev. J. Ireland Tucker, D.D..................... 1868
* Alexander L. Holley, LL.D........................... 1869–82
  Capt. Clarence E. Dutton, U. S. A.................... 1869–76
* Henry C. Lockwood................................ 1871–90
  William H. Doughty, C.E............................ 1871
* Hon. Thomas B. Carroll, *ex officio* Mayor.............. 1871–73
  Hon. Edward Murphy, Jr., "   "   "   1875–82
  Rev. William Irvin, D.D 1876
  John D. Van Buren, Jr., C.E.......................... 1876–82
  Charles Macdonald, C.E., LL.D................ 1880

James P. Wallace, C.E.................................... 1880
Joseph C. Platt, Jr., C.E............................... 1882
Elias P. Mann, C.E..................................... 1882
Hon. Edmund Fitzgerald, *ex officio* Mayor.............. 1882–86
Hon. Dennis J. Whelan,   "   "   "   1886–94
Stephen W. Barker, C.E................................. 1886
Henry B. Dauchy.. 1886
Henry G. Ludlow....................................... 1886
Robert W. Hunt.. 1886
Hon. John H. Peck, LL.D............................... 1887
Theodore Voorhees, C.E 1887
Edward C. Gale, C.E................................... 1887
John Squires, C.E...................................... 1888
Horace G. Young, C.E.................................. 1888
Paul Cook, A.M.. 1890
Hon. Francis J. Molloy, *ex officio* Mayor............... 1894

FACULTY AND INSTRUCTORS.

1824–1894.

EXECUTIVE OFFICERS OF THE FACULTY.

Senior Professors.

* Amos Eaton, A.M.. 1824–42
* George H. Cook, C.E., B.N.S............................ 1842–46
* Charles Drowne, C.E., A.M.............................. 1859–60

Directors.

B. Franklin Greene. C.E., A.M.......................... 1847–59
* Rev. N. S. S. Beman, D.D., LL.D....................... 1859–60
* Charles Drowne, C.E., A.M............................. 1860–76
William L. Adams, C.E.................................. 1876–78
David M. Greene, C.E.................................. 1878–91
Palmer C. Ricketts, C.E................................ 1892

PROFESSORS, INSRUCTORS, AND ASSISTANTS.

Astronomy.

* Charles Drowne, C.E., A.M., Professor.................. 1850–54
Dascom Greene, C.E.,   "   (Emeritus, 1893)... 1858–93
Charles W. Crockett, C.E.,   "   1893

Dascom Greene, C.E., Adjunct Professor............... 1856–58
Palmer C. Ricketts, C.E.,   Assistant Professor.......... 1882–84
Charles W. Crockett, C.E.,   "   "   1884–93
Palmer C. Ricketts, C.E.,   Assistant.................... 1875–82

Botany.

* Lewis C. Beck, M.D.,   Professor............... 1824–29
* John Wright, M.D.,   "   1836–46
* Frederick B. Leonard, M.D., "   1846–48
R. Halsted Ward, A.M., M.D.,   "   1869–92
Lewis G. Lowe, C.E., M.D.,   Lecturer............... 1855–56
* José Tell Ferrao, B.S.,   Repeater............... 1850–51
R. Halsted Ward, A.M., M.D., Instructor............... 1868–69
Edward R. Cary, C.E.,   "   1892

Chemistry.

* Amos Eaton, A.M.,   Professor.................. 1824–35
James Hall, A.M., LL.D.,   "   1835–41
* George H. Cook, C.E., B.N.S., "   1841–46
* William Elderhorst, M.D.,   "   1855–61
Charles A. Goessmann, Ph.D.   "   1861–64
Henry B. Nason,Ph.D., LL.D.,   •   1864
William P. Mason, C.E., M.D.,   "   1886
* William C. Bailey, B.N.S.,   Assistant Professor........ 1839–39
William P. Mason, C.E., M.D.,   "   "   1882–86
Jonathan R. Powell, C.E., Repeater.................... 1847–48
Lewis G. Lowe, C.E.,   "   1849–50
* José Tell Ferrao, B.S.,   "   1850–51
Dascom Greene, C.E.,   "   1852–53
James T. Allen, B.S.,   "   1854–55
Matthieu Darmstadt, Ph.D., Assistant.................. 1866–68
Irving A. Stearns, M.E.,   "   1868–69
* Edward Nichols, B.S.,   "   1871–73
Alfred S. Bertolet, M.E.,   "   1873–75
William P. Mason, C.E.,   "   1875–82

Civil Engineering.

* Amos Eaton, A.M.,   Professor............... 1828–42
* George H. Cook, C.E., B.N.S., "   1842–46
* Charles Drowne, C.E., A.M.,   "   1859–60
* William G. Lapham, C.E.,   Adjunct Professor 1838–39
* George H. Cook, C.E., B.N.S., "   "   1840–41

Descriptive Geometry and Drawing.

G. Gustavus Berger,	Professor................................	1850-51
S. Edward Warren, C.E.,	"	1853-72
Dwinel F. Thompson, B.S.,	"	1873
David Hathaway,	Instructor................................	1847-50
S. Edward Warren, C.E.,	"	1852-53
Dwinel F. Thompson, B.S.,	Assistant Professor..........	1872-73
Adolfo E. Besosa, C.E.	" "	1880-82
Albert H. Emery, C.E.,	Assistant....................	1855-58
William H. Powless, C.E.,	"	1875-76
* Herman Voorhees, C.E.,	"	1877-78
John A. L. Waddell, C.E.,	"	1878-80
Adolfo E. Besosa, C.E.	"	1880-82
Edgar B. Kay, C.E.,	"	1883-85
Robert A. Cairns, C.E.,	"	1885-87
James N. Ewing, C.E.,	"	1887-88
Edward F. Chillman, C.E.,	"	1888

English Language.

James T. Allen, B.S.,	Professor................................	1855-58
* T. Newton Willson, A.M.,	"	1859-59
* James R. Percy, C.E.,	Assistant....................	1857-59
Horace Loomis, C.E.,	Instructor....................	1862-65
Charles E. Illsley, A.B.,	"	1866-67
* Alexander G. Johnson, A.M.,	"	1869-75
John H. Kellom, A.M.,	"	1876-77
William W. Morrill, A.M.,	"	1877-82
Frank L. Nason, A.B.,	"	1882-88
John G. Murdoch, A.M.,	"	1888

French Language.

Louis Cousin, B.L.,	Professor................	1856-59
* Philip H. Baermann,	"	1862-66
* J. H. C. L. de Marcelleau, A.B.,	"	1869-73
Paul Edward Von Thun,	Instructor................	1852-54
George F. Struvé,	"	1854-56
John B. Luce, A.M.,	"	1860-61
* J. H. C. L. de Marcelleau, A.B.,	"	1866-69
* Jules Godeby, A.B.,	"	1873-90
Benedict Papot,	"	1891

Geodesy.

* Charles Drowne, C.E., A.M.,	Professor		1851-55
David M. Greene, C.E.,	"		1856-61
William H. Searles, C.E.,	"		1863-64
Charles McMillan, C.E.,	"		1865-71
William L. Adams, C.E.,	"		1872-78
David M. Greene, C.E.,	"		1878-91
William G. Raymond, C.E.,	"		1892
William H. Searles, C.E.,	Acting Professor		1862-63
William L. Adams, C.E.,	"	"	1864-65
Charles E. Smith, C.E.,	"	"	1871-72
* Thomas M. Cleeman, C.E.,	"	"	1891-92
William Fenton, C.E.,	Assistant Professor		1864-70
E. A. H. Allen, C.E.,	Repeater		1849-50
George B. Roberts, C.E., B.N.S.,	"		1850-51
William Tweeddale, C.E.,	Instructor		1852-54
* Joseph A. Moak, C.E.,	"		1854-55
David M. Greene, C.E.,	"		1855-56
* Joseph G. Fox, C.E.,	"		1861-62
William Fenton, C.E.,	"		1863-64
C. Whitman Boynton, C.E.,	Assistant		1856-57
Charles C. Martin, C.E.,	"		1856-57
William H. Powless, C.E.,	"		1875-76
* Herman Voorhees, C.E.,	"		1877-78
Robert R. Chadwick, C.E.,	"		1878-82
George R. Baucus, C.E.,	"		1882-84
John H. Emigh, C.E.,	"		1883
Harry L. Van Zile, C.E.,	"		1884-85
Charles W. Parks, C.E.,	"		1885-86
Augustus S. Kibbe, C.E.,	"		1886-87
John J. Berger, C.E.,	"		1887-88
Guy B. Waite, C.E.,	"		1888-90
Edward R. Cary, C.E.	"		1888

Geology.

* Amos Eaton, A.M.,	Professor		1824-42
* Ebenezer Emmons, A.M., M.D.,	"		1831-39
* George H. Cook, C.E., B.N.S.,	"		1842-46
Edward A. H. Allen, C.E.,	"		1850-54
James Hall, A.M., LL.D.,	" (Emeritus, 1876)...		1854-76
Robert P. Whitfield, A.M.,	"		1877-78
Henry B. Nason, Ph.D. LL.D.,	"		1878
Jonathan R. Powell, C.E.,	Repeater		1847-48

German Language.

*Philip H. Baermann, Professor........................ 1862-67
Paul Edward Von Thun, Instructor..................... 1850-54
George F. Struvé,   "   1854-56

Law of Contracts.

*James Forsyth, LL.D., Lecturer....................... 1875-86
John H. Peck, LL.D.,   "   1888

Mathematics.

B. Franklin Greene, C.E., A.M., Professor.............. 1847-50
*Charles Drowne, C.E., A.M.,   "   1850-55
Dascom Greene, C.E.,   Prof. (Emeritus, 1893).... 1858-93
Charles W. Crockett, C.E., A.M., " 1893
*Charles Drowne, C.E., A.M., Adjunct Professor......... 1849-50
Dascom Greene, C.E.,   "   "   1853-58
*T. Orlando Hopkins, C.E.,   Assistant Professor......... 1857-59
William Fenton, C.E.,   "   "   1864-70
Arthur W. Bower, C.E.,   "   "   1874-75
Palmer C. Ricketts, C.E.,   "   "   1882-84
Charles W. Crockett, C.E., A.M., "   "   1884-93
*Charles Drowne, C.E., A.M.,   Repeater.............. 1847-48
George W. Plympton, C.E.,   "   1850-50
George B. Roberts, C.E., B.N.S., "   1850-51
Dascom Greene, C.E.,   "   1852-53
De Volson Wood, C.E.,   Instructor.................. 1856-57
*Joseph G. Fox, C.E.,   "   1861-62
Horace Loomis, C.E.,   "   1862-64
William Fenton, C.E.,   "   1863-64
*George M. Hunt, C.E.,   "   1864-67
Arthur W. Bower, C.E.,   "   1871-74
Palmer C. Ricketts, C.E.,   Assistant................. 1875-82
Frank L. Nason, A.B.,   "   1882-88
John H. Emigh, C.E.,   "   1883
James M. Wilson, C.E.,   "   1885-86
George W. Worcester, B.S.,   "   1887-88
Guy B. Waite, C.E.,   "   1888-90
John G. Murdoch, A.M.,   "   1888
Daniel L. Turner, C.E.,   "   1891-92
James McGiffert, Jr., C.E.,   "   1892

Mechanics.

B. Franklin Greene, C.E., A.M.,   Professor.............. 1850–59
* Charles Drowne, C.E., A.M.,   "   1860–76
William H. Burr, C.E.,   "   1876–84
Palmer C. Ricketts, C.E.,   "   1884
* Charles Drowne, C.E., A.M., Adjunct Professor.......... 1850–51
William H. Burr, C.E.,   Assistant Professor............. 1876–76
Adolfo E. Besosa, C.E.,   "   "   1882–83
E. A. H. Allen, C.E.,   Repeater................ 1850–50
* James W. Bradshaw, C.E.,   "   1850–51
William Tweeddale, C.E.,   "   1852–54
George L. Moody,   "   1854–54
C. Whitman Boynton, C.E.,   "   1856–57
* T. Orlando Hopkins, C.E.,   "   1857–59
Arthur W. Bower, C.E.,   Instructor..................... 1871–75
William H. Burr, C.E.,   Assistant.................... 1875–76
William H. Powless, C.E.,   "   1877–78
John A. L. Waddell, C.E.,   "   1878–80
Adolfo E. Besosa, C.E.,   "   1880–82
George R. Baucus, C.E.,   "   1883–83
Guy H. Elmore, C.E.,   "   1883–84
William W. Cummings, C.E.,   "   1884–89
Hugh Anderson, C.E.,   "   1889

Mental Philosophy.

* N. S. S. Beman, D.D., LL.D., Lecturer................. 1841–54
* N. S. S. Beman, D.D., LL.D., Professor................ 1854–65

Metallurgy.

George W. Maynard, A.M.,   Professor................ 1867–71
Henry B. Nason, Ph.D., LL.D.,   "   1871

Natural History.

Edward A. H. Allen, C.E.,   Professor................ 1854–55
Henry B. Nason, Ph.D., LL.D.,   "   1858–64

Physics.

B. Franklin Greene, C.E., A.M., Professor... 1847–53
Charles A. Goessmann, Ph.D.,   "   1861–64
Arthur W. Bower, C.E.,   "   1878–80
Frank P. Whitman, A.M.,   '   1880–86
W. Le Conte Stevens, Ph.D.,   "   1892

Henry A. Rowland, C.E., Ph.D., Assistant Professor..... 1874–75
Arthur W. Bower, C.E.,   "   "   1875–78
Charles W. Parks, C.E.,   Acting Professor............. 1886–92
* Charles Drowne, C.E., A.M.,   Repeater................. 1847–50
Lewis G. Lowe, C.E., B.N.S.,   "   1850–50
* James W. Bradshaw, C.E.,   "   1850–51
William Tweeddale, C.E.,   "   1852–54
George L. Moody,   "   1854–55
Albert H. Gallatin, A.M., M.D.,   Lecturer............. 1866–67
Henry A. Rowland, C.E., Ph.D., Instructor............. 1872–74

Railroad Signals.

Pemberton Smith, C.E., Lecturer........................ 1892

Steam-engine.

David M. Greene, C.E.,   Professor................... 1378–91
H. de B. Parsons, B.S., M.E.,   "   1892
William J. Keep, C.E.,   Lecturer................. 1877–78

As shown in Chapter I, in the early days of the school, the teacher next in rank to the Senior Professor was called the Junior Professor, and the other instructors, who were appointed for a term or year, were called Assistants to the Senior Professor or to the Junior Professor.

Junior Professors.

* Lewis C. Beck, M.D...................................... 1824–29
* Hezekiah H. Eaton, A.B. (r.s.)........................... 1829–30
* Paul E. Stevenson, A.B. (r.s.)........................... 1830–31
* Ebenezer Emmons, A.M., M.D...................... 1831–39

Assistants to the Senior Professor.

* Fay Edgerton, A.B. (r.s.)................................ 1828
Thomas C. Ripley, A.B. (r.s.).......................... 1828
* Orlin Oatman, A.B. (r.s.)............................... 1829
* Daniel O. Comstock, A.B. (r.s.)....................... 1829
* James C. Booth... 1831
* S. Wells Williams, A.B. (r.s.)......................... 1832

D. Cady Smith, A.B. (r.s.)............................... 1833
* Alexander Van Rensselaer, A.B. (r.s.)................... 1833
* Theron R. Hopkins, A.B. (r.s.)......................... 1834
* Edward Suffern, C.E.................................... 1835
* Leman B. Garlinghouse, C.E............................. 1836
* George Johnson, C.E., B.N.S............................ 1836

Assistants to the Junior Professor.

* Timothy Dwight Eaton, A.B. (r.s.)...................... 1827
* Orlin Oatman, A.B. (r.s.)................. 1827
* John M. Barrows, A.B. (r.s.).......................... 1829
* Hezekiah H. Eaton, A.B. (r.s.)........................ 1829
* Douglas Houghton, A.B. (r.s.)......................... 1830
  James B. Dungan....................................... 1830
  Abel Storrs, A.B. (r.s.).............................. 1830
* Abram Sager, A.B. (r.s.)............................... 1831
  James Hall, A.B. (r.s.)............................... 1833

CATALOGUE OF GRADUATES.

1824–1894.

Name.	Degree.	Class.
Ackley, Calvin	C.E.	1854
Adam, Carl F	C.E.	1890
Adams, Edwin G	C.E.	1891
Adams, William L	C.E.	1862
*Addison, Alexander	C.E.	1866
Africa, James M	C.E.	1888
*Aguiar, A. W. F. de	C.E.	1867
Aguilera, Eugene M	C.E.	1887
Aguilera, Pedro T	C.E.	1887
Aguirregaviria, Casto	C.E.	1888
Aiken, William A	C.E.	1872
Albright, Joseph J	M.E.	1868
Alcover, Frederico M	C.E.	1871
Alden, John F	C.E.	1872
Aldrich, J. Franklin	C.E.	1877
Aldrich, Truman H	M.E.	1869
*Allaire, William M	C.E.	1876
Allen, Edward A. H	C.E.	1850
Allen, James T	B.S.	1855
Allen, Julian S	C.E.	1685
Allen, Kenneth	C.E.	1879
*Ambler, J. G	A.B. (r.s.)	1833
Amsden, Ik. E	C.E.	1891
Anderson, Hugh	C.E.	1886
Anderson, James C	C.E.	1876
*Anthony, C. H	B.N.S.	1840
Anzola, Roberto	C.E.	1869
*Appleton, Francis E	C.E.	1863
Appleton, Thomas	C.E.	1868
Arango, Ricardo M	C E.	1887
Argollo, Miguel de T	C.E.	1871
Arms, Edward W	C.E.	1869
*Arms, S. E	A.B. (r.s.)	1826
*Arnold, H	A.B. (r.s.)	1828
Arnold, John T	C.E.	1885
*Arnold, L. M	B.N.S.C.E.	1837
Arnold, William H	C.E.	1890
Ashby, Edward B	C.E.	1886
Arosemena, C.C	C.E., B.S.	1892
Auchincloss, W. S	C.E.	1862
Auerbach, Charles G	C.E.	1877
*Avery, Henry J	B.N.S.	1838
Aycrigg, William A	C.E.	1884
Babcock, Henry N	M.E.	1870
Babcock, W. Irving	C.E.	1878
Baermann, Palmer H	C.E.	1867
Bagg, Frederick A	C.E.	1893
Bagley, John A	C.E.	1853
*Baily, Joseph T	C.E.	1870
*Bailey, Thomas W	C.E.	1849
*Bailey, William C	B.N.S.	1838
Bainbridge, Francis H	C.E.	1884
Baker, Arthur G	C.E.	1876
Baker, Arthur L	C.E.	1873
*Baker, Henry	C.E.	1837
*Baker, William L	C.E.	1871
Balbin, Ernesto J	C.E.	1882
*Baldwin, William L	C.E.	1861
*Ball, Jasper N	C.E.	1848
Ball, R. Edward	C.E.	1875
Baltimore, Garnett, D	C.E.	1881
*Baltzell, Thomas K	C.E.	1854
Barber, Clarence M	C.E.	1878
Barcellos, J. J. A. de	B.S.C.E.	1868
Barker, Stephen W	M.E.	1868
Barnard, John F	C.E.	1850
Barney, Percy C	C.E.	1893
Barros, M. P. de	C.E.	1894
*Barrows, J. M	A.B.(r.s.)	1829
Bates, Frank C	C.E.	1889
Bates, William S	M.E.	1871
Baucus, George R	C.E.	1882
Baucus, William I	C.E.	1887
Baum, George	C.E.	1891
*Bayley, G. W. R	C.E.	1838
Beardsley, Arthur	C.E.	1867
Belding, Sherman W	C.E.	1891
*Bell, James E	C.E.	1873
Bement, Robert B. C	C.E.	1869
*Bement, R. B	A.B. (r.s.)	1830
*Benedict, Abner	A.B.(r.s.)	1826
Bergen, Van Brunt	C.E.	1863

Name.	Degree.	Class.
Berger, John J........	C.E.	1886
Bertolet, Alfred S.....	M.E.	1871
Besosa, Adolfo E......	C.E.	1875
Best, Arthur J........	C.E.	1877
*Billings, C. Jr........	C.E.	1877
Binsse, Henry L......	C.E.	1875
Birch, Charles E......	C.E.	1892
*Birdsall, James W...	T.E.	1860
Black, Alexander M...	C.E.	1869
Blaisdell, Anthony H..	C.E.	1870
Blake, J. J...........	C.E.	1894
Blandy, Isaac C.......	C.E.	1887
Blanton, L. Harvie....	C.E.	1877
Bloss, Jabez P......	B.N.S.	1846
Bloss, Richard P......	C.E.	1881
*Blun, Abraham......	C.E.	1873
Boardman, Arthur E..	C.E.	1870
Boardman, Henry M..	C.E.	1871
Bogue, Virgil G.......	C.E.	1868
Boller, Alfred P......	C.E.	1861
Boller, Frederick J...	C.E.	1869
Bontecou, Reed B..	B.N.S.	1842
*Booth, James C.[1]....	Ph.D.	1831
Bostrom, Augustus O..	C.E.	1877
Botero, Fabriciano....	C.E.	1885
Bours, Benjamin W...	C.E.	1839
Bowen, Franklin H....	C.E.	1883
Bower, Arthur W.....	C.E.	1871
Boyd, William H..A.B.(r.s.)		1832
Boynton, C. Whitman..	C.E.	1856
*Bradshaw, James W..	C.E.	1850
Bradway, J. R..C.E.,B.N.S.		1841
*Brainerd, George B..	C.E.	1865
Breese, James L......	C.E.	1875
Breithaupt, William H..	C.E.	1881
Bridgers, Robt. R. Jr...	C.E.	1879
*Briggs, Caleb..B.N.S.,C.E.		1835
Briggs, Roswell E....	C.E.	1868
*Brinley, E. Jr..B.N.S.,C.E.		1839
Brinsmade, Henry N...	C.E.	1879
*Brodt, J. H...C.E.,B.N.S.		1844
Brown, Marshall W...	C.E.	1894
Brown, N. W. L.......	C.E.	1892
Brown, Robert K......	C.E.	1888
Brown, Thurber A....	C.E.	1883
*Browne, Percy T.....	C.E.	1863
Bruckman, T. G......	B.S.	1890
*Bryant, Cyrus..A.B. (r.s.)		1829
*Bryant, Fred M......	C.E.	1873
*Buck, B. Franklin....	C.E.	1837
Buck, Leffert L........	C.E.	1868
Buck, Richard S., Jr..	C.E.	1887
*Buckhout, Nathan W..	C.E.	1862

Name.	Degree.	Class.
*Buckingham, E. P....	C.E.	1861
Bucknell, Elmer J.....	C.E.	1892
Buel, Albert W.......	C.E.	1883
*Buel, Richard H.....	C.E.	1862
Buel, Samuel, Jr......	C.E.	1865
*Bullard, G.....	A.B. (r.s.)	1828
Burden, Henry.......	M.E.	1869
Burden, James........	C.E.	1892
Burdett, Edward A....	C.E.	1876
Burge, Alfred W......	C.E.	1893
*Burgess, William N..	M.E.	1869
Burhans, Frederic O..	B.S.	1853
*Burnett, Leicester....	C.E.	1856
Burnham, George, Jr..	C.E.	1872
Burr, William H......	C.E.	1872
*Burrall, William H...	C.E.	1851
Bushnell, Joseph, Jr...	C.E.	1877
*Buswell, E.G..B.N.S.,C.E.		1841
Butler, Lawrence P....	C.E.	1890
Butt, McCoskry.......	C.E.	1882
Buxton, Clifford......	C.E.	1865
*Byram, William H...	C.E.	1877
Cabot, William B.....	C.E.	1881
Cains, Robert A......	C.E.	1885
Caldwell, Charles A...	C.E.	1888
Caldwell, James H....	B.S.	1886
Caldwell, James N.,Jr..	C.E.	1874
Callery, William V....	C.E.	1886
Campbell, Charles.....	C.E.	1873
Campbell, Charles W..	C.E.	1879
*Campbell, James...B.N.S.		1843
Campbell, Joseph H..	M.E.	1868
Cantanhede, P. de C...	C.E.	1881
Carbonell, Carlos F....	C.E.	1875
Card, William D.......	C.E.	1890
Carnrick, George W...	C.E.	1874
Carr, Ezra S...C.E.,B.N.S.		1838
Carter, Edward C.....	C.E.	1876
Cary, Edward R......	C.E.	1888
Casanova, José N.....	B.S.	1859
Cassatt, Alexander J..	C.E.	1859
Castro, Alberto de....	T.E.	1860
Ceballos, G. F. de.....	C.E.	1868
Chadwick, Robert R..	C.E.	1878
*Chamberlaine, N. H..	C.E.	1856
Chambers, Frank T...	C.E.	1892
Cahmbers, John.......	C.E.	1886
Chambers, John S.....	C.E.	1881
Chambers, Ralph H...	C.E.	1893
*Chandler, J....A.B. (r.s.)		1827
Chesrown, Elias......	C.E.	1885
Chibas, Eduardo J.. ..	C.E.	1889
Chillman, Edward F..	C.E.	1888

Name.	Degree.	Class.
*Chislett, John J.	C.E.	1884
Chrysler, Frank	C.E.	1884
*Chubb, A. L.	C.E.,B.N.S.	1848
Church, Daniel W.	C.E.	1877
Church, Frederick B.	C.E.	1891
Church, Townsend V.	C.E.	1881
Church, W. Lee	C.E.	1872
Cintra, Francisco de A.	C.E.	1881
Clark, Dorlon	C.E.	1885
Clark, Frank L.	C.E.	1880
Clark, Joseph E.	B.N.S.	1845
Clark, John A.	C.E.	1887
Clark, John M.	C.E.	1856
*Clarke, Jos. B.	A.B. (r.s.)	1829
*Cleemann, Thos. M.	C.E.	1865
*Clement, William H.	C.E.	1835
Clinch, J. Morton	C.E.	1854
Cobb, Arthur	C.E.	1880
*Cobb, James C.	A.B. (r.s.)	1831
Cogswell, William B.[1]	C.E.	1851
Coit, James C	C.E.	1858
Colby, Archie L.	C.E.	1887
Colby, John D	C.E.	1884
Colby, S. K	C.E.	1894
Collin, D., Jr.	C.E.,B.N.S.	1842
Collingwood, Francis.	C.E.	1855
*Collins, Chas.	C.E.,B.N.S.	1840
*Comstock, D.O.	A.B.(r.s.)	1829
Connett, Albert N.	C.E.	1880
Converse, Wade.	C.E.	1880
*Cook, Albert B.	C.E.	1892
*Cook, Charles R.	C.E.	1837
*Cook, Geo. H.	C.E.,B.N.S.	1839
*Cook, Robert G.	B.N.S.	1847
Cooley, Lyman E.	C.E.	1874
Cooper, Theodore.	C.E.	1858
*Cotes, Elihu W.	C.E.	1839
*Cotterell, Nathan	C.E.	1841
Cottman, Joseph B.	B.N.S.	1835
Coulson, Benjamin L.	C.E.	1893
Courtenay, William H.	C.E.	1879
Covode, James H.	C.E.	1882
Cox, Abraham B., Jr.	C.E.	1867
Cox, Leonard M.	C.E.	1892
Craft, Charles C.	C.E.	1866
Crafts, Walter.	C.E.	1859
Craig, Washington R.	C.E.	1893
Cramer, Eliphalet W.	C.E.	1879
*Crehore, C. F.	C.E.	1848
*Crocker, E. B.	A.B.(r.s.)	1833
Crockett, Charles W.	C.E.	1884
*Cromwell, James.	C.E.	1861
Crosby, Homer.	C.E.	1887

Name.	Degree.	Class.
Crosby, Horace	C.E.	1862
Crosby, Wilson	C.E.	1856
*Cross, Charles E.	C.E.	1855
*Cross, Charles S.	C.E.	1838
Cummings, C. A.,	B.N.S.,C.E.	1849
Cummings, Fred. M.	C.E.	1886
Cummings, W. W.	C.E.	1884
Cunningham, A. C.	C.E.	1885
Cunningham, Seymour.	C.E.	1884
Cuntz, Johannes H.	C.E.	1886
Curfman, S.B.	C.E.	1894
Curtis, Henry	C.E.	1854
Curtis, John H.	C.E.	1873
Cushman, George H.	C.E.	1879
Dabney, Frederic Y.	C.E.	1857
Danforth, Henry W.	C.E.	1842
*Danker, Albert.	A.B.(r.s.)	1826
Dauchy, Edward N.	C.E.	1840
Dauchy, Walter E.	C.E.	1875
Davenport, Ezekiel C.	C.E.	1886
Davenport, Fred.	C.E.	1892
Davenport, Henry B.	C.E.	1886
*Davey, John J.	A.B.(r.s.)	1827
Davis, Charles H.	C.E.	1884
Davis, Chester B.	C.E.	1877
Davis, Joseph P.	C.E.	1856
Davis, Josiah R. T.	C.E.	1876
Davison, George S.	C.E.	1878
Deal, Alvin E.	C.E.	1882
Deal, Elvin A.	C.E.	1882
*Decker, T. W.	A.B.(r.s.)	1830
De Leon, Moise.	C.E.	1892
Denegre, William P.	C.E.	1877
*Dennis, George R.	B.N.S.	1839
Denny, Stacey E.	C.E.	1891
*Devol, Edward.	A.B.(r.s.)	1831
Dias, Eduardo da R.	C.E.	1891
Dias, Luiz da R., Jr.,[2]	T.E.,C.E.	1860
Diehl, George C.	C.E.	1894
Dike, Albyn P.	C.E.	1877
Dodge, Richard D.	C.E.	1860
Doughty, William H.	C.E.	1858
Drake, Tracy C.	B.S.	1886
*Drayton, Henry J.	C.E.	1839
*Drayton, James S.	B.N.S.	1836
*Drew, Francis G.	A.B.(r.s.)	1827
*Drowne, Charles,	C.E.,B.N.S.	1847
Duane, Harry B.	C.E.	1878
Duane, James.	C.E.	1873
Durbin, James G.	C.E.	1884

[1] Degree conferred, 1884.

[2] Degree C.E. conferred, 1882.

Name.	Degree.	Class.
*Durham, Anson	C.E.	1840
Earle, Thomas	C.E.	1887
Easby, Marmaduke W.	C.E.	1886
Easby, Paul H	C.E.	1886
Easby, William, Jr	C.E.	1890
*Eaton, H. H.	A.B. (r.s.)	1826
*Eaton, T. D.	A.B. (r.s.)	1826
Echeverria, Juan F.	C.E.	1885
Eckert, Edward W	C.E.	1875
*Eddy, Jacob F	C.E.	1835
*Edgerton, Fay	A.B. (r.s.)	1828
Edwards, John C	C E.	1885
Edwards, R.	B.N.S.,	1848
Edwards, Thomas H.	C.E.	1891
Eguiguren,Vincente F.	C.E.	1888
Elder, George R	C.E.	1884
Eldridge, Archibald R.	C.E.	1888
Eldridge, Griffith M	C.E.	1885
Ellis, George E	C.E.	1892
Ells, George F	C.E.	1856
Elmer, Howard N	C.E.	1877
Elmore, Guy H	C.E.	1883
Ely, Theodore N	C.E.	1866
Emerson, Rufus H	C.E.	1861
Emery, Albert H	C.E.	1858
Emigh, John H	C.E.	1879
Emmerich, Edward E.	C.E.	1892
*Emmons, E.	A.B. (r.s.)	1826
Emory, Gustavus W.	C.E.	1887
*Emory,Thomas.	A.B.(r.s.)	1828
Endicott, Mordecai T.	C.E.	1868
Endress, William F.	C.E.	1879
Ensign, Milton W	C.E.	1871
Eppele, Frank J	C.E.	1888
Escarza, Sotera E	C.E.	1894
*Escobar, José	C.E.	1867
Escobar, Roberto	C.E.	1857
Estabrook, John D	C.E.	1856
Estep, Josiah M., Jr	C.E.	1889
Eva, Samuel J	C.E.	1891
*Evans, W.W	C.E.	1836
Ewens, John	C.E.	1878
Ewing, J. Nelson	C.E.	1887
Fabian, William J	C.E.	1874
Fairchild, Henry M	C.E.	1886
Farnum, Henry H	C.E.	1865
Farwell, Elmer S	C.E.	1891
*Fay, Francis F	M.E.	1868
Feldmeier, Harvey	C.E.	1892
Felton, Herbert C	C.E.	1866
Fenton, William	C.E.	1861
*Ferrao, José Tell	B.S.	1850
Ferris, G. W. G., Jr	C.E.	1881

Name.	Degree.	Class.
*Ferriss, John A., Jr	C.E.	1875
Fickes, Edwin S	C.E.	1894
*Field, Charles S	C.E.	1838
Fields, Samuel J	C.E.	1867
Filer, Walter G	C.E.	1890
*Fish, Dean	B.S.	1886
*Fisher, Charles H.[1]	C.E.	1853
Fisher, Clark	C.E.	1858
*Fisher, Joseph S	C.E.	1849
Fisher, Tucker H	C.E.	1875
*Fitch, Asa, Jr.	A.B. (r. s.)	1827
Flynn, John, Jr	C.E.	1894
*Follin, Ormond W.	B.S.	1859
Ford, Edwin	C.E.	1882
Ford, Frank L	C.E.	1874
*Ford, Jo. ' Ç. A	C.E.	1866
Forsyth, Robert	C.E.	1869
Fortun, y Andre S	C.E.	1889
Foster, Albert W	C.E.	1871
Foster, Thomas J	C.E.	1892
Fowler, Albert C	C.E.	1878
Fowler, Clarence A	C.E.	1885
*Fox, Albert R.	A.B. (r.s.)	1830
*Fox, Joseph G	C.E.	1861
Fox, Peter H	C.E.	1864
Fox, S. Waters	C.E.	1876
*Fox, William L	C.E.	1875
Franco, Antonio de B.	C.E.	1890
Franco, Eugenio de L.	C.E.	1878
Frank, Isaac W	C.E.	1876
Frazier, J. W	C.E.	1894
Freeman, Ernest G	C.E.	1888
Freeman, Harold A	C.E.	1876
Fritcher, George E.	C.E.	1878
Frith, Arthur J	C.E.	1873
Frothingham, J. H.		
	B.N.S., C.E.	1849
Fuertes, Estévan A	C.E.	1861
Fuess, F. F	C.E.	1894
Gale, E. Courtland	C.E.	1883
*Gale, E. Thompson	C.E.	1837
*Gale, George A	B.N.S.	1847
*Garcia, F. Garcia y.	C.E.	1872
Gardner, Arthur B.	C.E.	1891
Garland, W. S	C.E.	1894
*Garlinghouse, L. B	C.E.	1837
Garlinghouse,Fred. L.	C.E.	1871
Garzon, Julio N	C.E.	1894
Gasteazoro, Carlos A.	C.E.	1891
*Gearn, Walter A	C.E.	1878
Geer, Harvey M	C.E.	1872
Gest, Alexander P	C.E.	1874
Geuder, Edward G	C.E.	1876

[1] Degree conferred, 1882.

Name.	Degree.	Class.
Gibbs, L.A..C.E. 1892; B.S.		1893
Giberga, Ovidio.......C.E.		1885
*Giblin, Arthur L.....C.E.		1891
Gifford, George E.....C.E.		1887
Give, Henry De.......C.E.		1888
Goetzmann, F. G......C.E.		1878
Goicouria, A. V. de....C.E.		1871
*Gold, Miner....A.B. (r.s.)		1829
Goldstein, Max L.....C.E.		1867
Gonzalez, Juan, Jr....C.E.		1870
Gottlieb, R. D.......C.E.		1890
*Gould, James P......C.E.		1863
Gowing, Burdett C....C.E.		1861
Grant, Bertrand E....C.E.		1890
*Grant, Edward M....C.E.		1860
Gray, John H........C.E.		1887
Greeley, Samuel S....C.E.		1846
Green, Lansdale B....C.E.		1891
Greenalch, Wallace...C.E.		1893
Greene, Albert S.....C.E.		1859
Greene, B. F..C.E., B.N.S.		1842
Greene, Dascom......C.E.		1853
Greene, David M.....C.E.		1851
*Greene, George M....C.E.		1859
Greene, Joseph S.....C.E.		1878
Gregory, Brainerd E...C.E.		1887
Gridley,V.H. B.S.1893;C.E.		1894
*Griffen, George S.....C.E.		1874
Griffen, Henry R......C.E.		1877
Griffith, Charles G....C.E.		1877
Grimes, Charles L.....C.E.		1871
Grinnell, Frederic.....C.E.		1855
Griswold, John W......B.S.		1865
Groesbeck, Geo. S.....C.E.		1889
Gronau, William F....C.E.		1887
Grove, Independence..C.E.		1882
Guerra, Arturo........C.E.		1876
Guerrero, Cárlos..B.S.C.E.		1867
Gunn, Frederick C....C.E.		1887
Gurley, Lewis E......C.E.		1845
Gurley, Louis W......C.E.		1882
*Gurley, William.....C.E.		1839
*Haddock, Arba R....C.E.		1862
Haight, Theodore S...C.E.		1885
Hailman, James D....C.E.		1887
Hall, Fitz Edward....C.E.		1842
*Hall, George M......C.E.		1849
*Hall, G. Thomas.....C.E.		1868
Hall, James.....A.B. (r.s.)		1832
Hall, John G.........B.S.		1887
*Hall,William.C.E., B.N.S.		1846
Hallock, James C.....C.E.		1891
Hallsted, James C.,Jr..C.E.		1883
Hammond, W. B......C.E.		1880
Haraguchi, Kaname...C.E.		1878

Name.	Degree.	Class.
Harison, R. Morley...C.E.		1879
*Harley, Henry.......C.E.		1858
*Harper, Albert M....C.E.		1867
Harris, Charles P.....C.E.		1873
*Harris, Henrique....T.E.		1860
*Harris, Joel B.......C.E.		1841
Harris, William P.....C.E.		1866
Harrison, Frank......C.E.		1888
Harrold, Thomas, Jr..C.E.		1887
*Haskell, S.E..B.N.S.,C.E.		1845
Haskin,A. N...B.N.S.,C.E.		1840
*Haskin, A. B..C.E.,B.N.S.		1840
Haskin, Leonard W...C.E.		1841
Haskin, William L....C.E.		1861
Hassinger,William H..C.E.		1885
Hauck, Albert L......C.E.		1886
Hawley, Wm. C......C.E.		1886
*Hawley, F. J.C.E., B.N.S.		1837
Hayt, Stephen T......C.E.		1882
Hearne, Frank J......C.E.		1867
Hébert, Paul O.......C.E.		1889
Hedden, Eugene B....C.E.		1885
Heizmann,Theodore I..C.E.		1859
Henderson, William...C.E.		1876A
Henry, John J........C.E.		1881
Henry, Philip W......C.E.		1887
Henry, Wm. G....A.B.(r.s.)		1828
Hepburn, Fred T......C.E.		1893
Hermann, Edward A..C.E.		1879
Hernandez, José......C.E.		1867
Hetzel, James........C.E.		1885
Hewes, Virgil H......C.E.		1881
Heyl, Jacob E........B.S.		1870
Hill,AugustusG..A.B.(r.s.)		1831
Hilt, F. K............C.E.		1894
Himmelwright, A.L.A.C.E.		1888
*Hinckley, Frank.....C.E.		1863
Hine, Samuel K.......B.S.		1892
Hinsdale,TheodoreR..C.E.		1886
Hirai, Seijiro.........C.E.		1878
Hitchcock. Dwight A..C.E.		1886
Hoadly, Edward M....C.E.		1889
Hodge, Harry S......C.E.		1878
Hodge, Henry W......C.E.		1885
Hoeing, Joseph B.....C.E.		1876
Holmes, Henry.......C.E.		1855
*Holton, George C....C.E.		1860
Hood, Richard H......C.E.		1887
Hopkins, Albert L.....C.E.		1892
Hopkins, James B.....C.E.		1886
*Hopkins, T. Orlando.C.E.		1857
*Hopkins, T. R..A.B. (r.s.)		1834
Horbach, Paul W......C.E.		1886
*Horsford, Eben N....C.E.		1838
*Horton, G. F...A.B. (r.s.)		1827

Name.	Degree.	Class.
Horton, George T.....C.E.		1893
*Horton, J. S....A.B. (r.s.)		1829
*Houghton, D...A.B. (r.s.)		1829
Houghton, J. F........C.E.		1848
House, S. R.....A.B. (r.s.)		1834
Howard, James W....C.E.		1888
*Howard, Jerome B...C.E.		1838
Hubbell, George S....C.E.		1886
*Hulbert, A.....A.B. (r.s.)		1826
*Humphreys, John G..C.E.		1873
Humphrey, Henry C..C.E.		1887
Hunt, Conway B......C.E.		1882
*Hunt, George........C.E.		1858
*Hunt, George M.....C.E.		1866
Huntington, W. W....C.E.		1876
Huntley, Lay.........C.E.		1876
Hurd, Tyrus W.......C.E.		1836
Hutton, Frank C......C.E.		1885
Hyde, Charles B......C.E.		1841
Hyde, Douglass W....C.E.		1841
Illsley, Charles E......C.E.		1867
Ingham,W. A..C.E.,B.N.S.		1846
*Ishigro, Taro........C.E.		1881
Jack, E. E. B.........C.E.		1894
*Jackson, S. C...A.B. (r.s.)		1827
Jaggard, Herbert A...C.E.		1889
Jarrett, Edwin S......C.E.		1889
Jeffers, William W....C.E.		1892
Jenkins, Lewis L......C.E.		1882
*Jennings, Henry C...C.E.		1879
*Jenny, Joseph H.....C.E.		1841
Jewett, Charles H.....C.E.		1885
*Johnson, G...C.E., B.N.S.		1837
Johnson, I. G..B.N.S., C.E.		1848
Johnson, James M.....C.E.		1879
Johnson, Stewart.....C.E.		1887
Johnston, Thomas T..C.E.		1877
Jones, Walter S.......C.E.		1893
*Judson, Charles T....C.E.		1875
Just, George A........C.E.		1881
Kaufman, Gustave....C.E.		1880
Kay, Edgar B.........C.E.		1883
Kay, William G.......C.E.		1875
Keenan, John J.......C.E.		1888
Keeney, J. C....A.B. (r.s.)		1827
*Kellogg, E. R.C.E.,B.N.S.		1841
Kellogg, Nathan......C.E.		1841
Kellogg, Norman B...C.E.		1873
Kellogg, Warren T....C.E.		1861
Kelly, John P.........C.E.		1876
*Kendall, David......C.E.		1838
Kibbe, Augustus S....C.E.		1886
Kiersted,Wynkoop,Jr..C.E.		1880
Kilbourne,Edward W..C.E.		1885
Kimball,FrederickN...C.E.		1886

Name.	Degree.	Class.
Kimberly, John A.,Jr..C.E.		1889
*King, William J.......C.E.		1880
*Kingman, L. H..A.B.(r.s.)		1829
Kingsley, James C....C.E.		1876
Kirby, G. Frederic....C.E.		1857
Kirtland, Alfred P....C.E.		1871
Kirtland, E. F........C.E.		1894
Knap, Joseph M......C.E.		1858
*Knap, Thomas L.....C.E.		1866
Knapp, George O......C.E.		1876
*Kneass, Strickland...C.E.		1839
Kneass, Strickland, L..C.E.		1880
*Knickerbacker,H.Jr..C.E.		1887
Knickerbacker, John..C.E.		1886
Knowlton,Theodore E.C.E.		1893
*Krause, Conrad B....C.E.		1879
Kummer, Fred. A.....C.E.		1894
Lacerda, Augusto de..B.S.		1855
La Chicotte, Henry A..C.E.		1885
La Coste, Louis.......C.E.		1841
Laflin, Louis E........C.E.		1882
*Lally, James.........C.E.		1861
Landor, Edward J.....C.E.		1876
Lane, Edward V. Z....C.E.		1875
Lapeyre, James M....C.E.		1892
*Lapham, William G..C.E.		1838
Lavandeira, Antonio..C.E.		1877
Lawlor, Joseph M.....C.E.		1888
Lawlor, Thomas F....C.E.		1886
*Lawrance, B. R......B.S.		1868
Lawton, Frederick B..C.E.		1891
Lay, Henry C.........C.E.		1875
Lea, George H........C.E.		1872
Leme, Luiz G. da S....C.E.		1880
Lempe, Fred J.........C.E.		1893
*Lent, George B.......C.E.		1838
Lesley, Alex. M....B.N.S.		1846
Leverich, Gabriel.....C.E.		1857
Lewis, Nelson P......C.E.		1879
*Lewis, Wm..C.E., B.N.S.		1840
*Lilienthal, Benj. N...C.E.		1866
*Lindsley, Aaron L....C.E.		1842
Lippincott, Jason E....C.E.		1883
Llano, Antonio........C.E.		1890
Locke, Elmer H....B.N.S.		1848
Lockhart, John M.....C.E.		1887
*Lockling,L.L..C.E.,B.N.S.		1837
Long, Thomas J.......C.E.		1873
*Loomis, Charles L...C.E.		1851
Loomis, Horace......C.E.		1865
Low, Samuel B........C.E.		1876
Lowe, Jesse...........C.E.		1885
Lowe, L. G....C.E., B.N.S.		1848
Lowrey, G....C.E., B.N.S.		1845
Luaces, Ernesto L.....C.E.		1867

Name.	Degree.	Class.
Ludwig, Julius A	C.E.	1889
*Mabbett, H. J..A.B. (r.s.)		1833
Macdonald, Charles	C.E.	1857
Macfarlane,Graham.	C.E.	1872
MacGregor,George C..C.E.		1871
Macksey, Henry V	C.E.	1886
Mader, Arthur B	C.E.	1890
Magor, Henry B	C.E.	1894
Maguire, J. B	C.E.	1894
Mallory, George B...	C.E.	1867
Mallory, Marshall H..C.E.		1865
Man, Albon P	C.E.	1866
Mann, Elias P	C.E.	1872
*Mann, George H	C.E.	1870
Mansfield, M. William.C.E.		1871
Manville, C. Rollin	C.E.	1880
Marburg, Edgar	C.E.	1885
Marcy, William	B.S.	1893
*Marks, J. Harrod	C.E.	1871
Marlett, S. H..C.E., B.N.S.		1841
Marling, William	C.E.	1872
Marshall, Thomas F...C.E.		1867
Marstrand, O. Julius..C.E.		1882
Martin, Charles C	C.E.	1856
Martin, John L	C.E.	1894
Martin, William H	C.E.	1856
Martins, José C	C.E.	1886
Mason, William P	C.E.	1874
Masses, Jose D	C.E.	1882
Masten, Cornelius S	C.E.	1850
Matas, Ramon	T.E.	1860
*Mather, Charles R...M.E.		1870
Matsmoto, Souichiro..C.E.		1876
Matthews, Irving E	C.E.	1887
Mauldin, Thomas S	C.E.	1891
*Maxwell, William B..C.E.		1875
May, John E..B.N.S., C.E.		1846
McCartney, W. M	C.E.	1894
McCaughin, John	C.E.	1842
*McClellan, Henry G..C.E.		1869
McClelland, Wilson	C.E.	1886
McClintock, Hugh P..C.E.		1880
McComb, Edward C...C.E.		1887
McCord, William S	C.E.	1881
McGiffert, James, Jr...C.E.		1891
McGuire, James C	C.E.	1888
McHarg, Arthur V. A..C.E.		1892
McKay, George A	C.E.	1894
McKee, Aaron G	C.E.	1836
*McKee, Robert G	C.E.	1835
McKinney, Samuel P..C.E.		1884
McKnew, William H..C.E.		1878
McLaren, Daniel	C.E.	1878
McLean, John	C.E.	1876

Name.	Degree.	Class.
*McManus, P.C.W.,		
A.B. (r.s.)		1826
McMillan, Charles	C.E.	1860
McNaugher,David W..C.E.		1885
*McNeill, Elmore B...C.E.		1881
McPherson, J. A	B.S.	1894
Megear, Alter	C.E.	1868
Melchert, A. C. d'A...C.E.		1886
Mendoza, Victor G. de.C.E.		1888
Menocal,AnicetoG. de.C.E.		1862
Menocal, Arturio N...C.E.		1881
*Merian, Henry W	C.E.	1858
*Merrifield, Paul S	C.E.	1878
*Metcalf, J. B...A.B. (r.s.)		1829
Metcalf, William	C.E.	1858
Miller, Leverett S	C.E.	1885
*Miller, S. V. R	C.E.	1841
*Millet, Albert H	C.E.	1867
Mills, Hiram F	C.E.	1856
Mills, William W	C.E.	1868
Mitchell, Horace H	C.E.	1887
*Moak, Joseph A	C.E.	1854
Molina, Ricardo V	C.E.	1887
Moliner, Julio S	C.E.	1888
Montgomery, Neil R..C.E.		1885
Montony, Liberty G...C.E.		1890
Moore, Frank L	C.E.	1867
Moore, Marshall G	C.E.	1884
Morris, Thos. O'Neil..C.E.		1870
Morse, Henry G	C.E.	1871
Morton, Nathaniel	C.E.	1850
Moss, Charles H	C.E.	1867
*Mullin, A. T. E	C.E.	1861
Mullin, Joseph	C.E.	1869
Munoz del Monte,A.C..C.E.		1886
Munoz del Monte,L.E..C.E.		1888
*Murphy,J.W..B.N.S.,C.E.		1847
Murray, Jo. Dorr	C.E.	1892
Myers, John H	C.E.	1893
Mynderse, Edward	C.E.	1838
Naranjo, Francisco R. C.	B.S.	1863
Neal, Robert C	M.E.	1870
Neilson, Robert	C.E.	1861
Nelles, George T	C.E.	1877
Nellis, Dan H	C.E.	1892
Newbold, Thomas E..C.E.		1882
*Nichols, Edward	B.S.	1871
Nichols, Othniel F	C.E.	1868
Nicholson,William A...C.E.		1877
Nickel, George D	M.E.	1870
Nickerson, J. G	B.N.S.	1848
Nier, John W	C.E.	1876
Norris, Aleck J	C.E.	1886
Nugent, Paul C	C.E.	1892

Name.	Degree.	Class.
*Oakey, James	C.E.	1837
*Oatman, Orlin	A.B.(r.s.)	1827
*Olmstead, A.B.,	C.E.,B.N.S.	1837
Olmstead, H. L.	C.E.	1894
*Olmstead, L.G.	A.B.(r.s.)	1830
*Olyphant, H. V.	C.E.	1868
Osborn, Frank C.	C.E.	1880
*Osborn, G. K.	A. B.(r.s.)	1830
Osborne, Charles M.	C.E.	1853
*Ostrom, John	C.E.	1857
Otto, John B.	C.E.	1871
Packard, Ralph G.	C.E.	1664
Painter, A. E. W	C.E.	1863
Painter, Edward L.	C.E.	1884
Painter, Herbert B.	C.E.	1891
Palmer, Miguel C.	C.E.	1894
Pardee, Ario, Jr.	C.E.	1858
Pardee, Calvin	B.S.	1860
Parish, Wainwright	C.E.	1888
*Park, A. F.	C.E.,B.N.S	1840
Parker, Charles M.	C.E.	1889
Parks, Albert F.	C.E.	1891
Parks, Charles W.	C.E.	1884
*Parkinson, John B.	C.E.	1876
Parrish, Edward, Jr.	C.E.	1870
*Parsons, Samuel B.	C.E.	1840
Patten, Henry B.	C.E.	1878
*Paterson, S. V. R.	C.E.	1836
Pattison, Harry D.	C.E.	1874
*Pearce, Allen	C.E.	1838
Pearl, James W.	C.E.	1880
Pease, Charles S.	C.E.	1876
*Peck, Hollam L.	C.E.	1849
Peck, William A.	C.E.	1869
Peebles, Robert C.	C.E.	1869
Pelaez, Manuel A.	C.E.	1873
*Pelton,Wm. S.	A.B.(r.s.)	1826
Pemberton, John, Jr.	C.E.	1860
Penfield, James A.	B.N.S.	1846
*Percy, James R.	C.E.	1859
Perkins, Charles P.	C.E.	1866
Perry, Thornton T.	C.E.	1885
Peterson, B. Walker.	C.E.	1877
Pettibone, C.V.	C.E.	1867
*Philip, John H.	A.B.(r.s.)	1832
*Philip, J. V. N.	B.N.S.	1839
Pierce, George H.	C.E.	1858
*Pierpont, John	M.E.	1869
*Pike, Samuel J.	A.B.(r.s.)	1830
Pirajá. J. R. da S., Jr.	C.E.	1865
Platt, Frank E.	C.E.	1879
Platt, Joseph C.	C.E.	1866
*Platt, Merritt	A.B. (r.s.)	1830
Plumb, James Ives.	C.E.	1886

Name.	Degree.	Class.
Plummer, George C.	C.E.	1892
Plympton, George W.	C.E.	1847
Pomeroy, Halsey B.	C.E.	1867
*Pomeroy, H.	C.E., B.N.S.	1841
Pond, Frank	C.E.	1875
*Post, James H	C.E.	1839
*Potter, Charles F.	C.E.	1878
*Potter, Clarkson N.	C.E.	1843
Potter, Henry W.	C.E.	1879
*Potter, George C.	C.E.	1839
Potter, Winfield S.	C.E.	1890
Potts, Benjamin C.	C.E.	1863
Powell, Ambrose V.	C.E.	1868
Powell, J. R.	C.E., B.N.S.	1846
Powell, Marcus	C.E.	1889
Powell, William J.	C.E.	1839
Powers, Joseph A.	C.E.	1880
Powless, William H.	C.E.	1874
*Pratt, Charles S.	C.E.	1883
*Pratt, Ira R.	C.E., B.N.S.	1842
Pratt, Robt. J.	C.E.	1883
Pratt, William M	C.E.	1857
*Prescott, Richard	M.E.	1871
Price, Victor T.	C.E.	1888
*Prime, A. J.	A.B.(r.s.)	1829
*Pruyne, DeForest	C.E.	1876
*Putnam, George	C.E.	1838
Quackenbush,John H.	C.E.	1856
Quintana, Manuel P.	C.E.	1884
Rae, Charles W.	C.E.	1866
Raht, Adolphus W.	C.E.	1877
Rainsford, Thos. H.	C.E.	1881
Ralston, John B.	C.E.	1888
Randolph, John H.	C.E.	1870
Ranney, Marcus H.	C.E.	1868
Ranney, Willet G.	C.E.	1890
Raymond, Charles T.	C.E.	1879
Raymond, Thomas C.	C.E.	1865
Raynolds, James D.	C.E.	1870
Reed, James	C.E.	1873
Reed, Paul L.	C.E.	1894
Reeves, David	C.E.	1872
Reeves, William H.	C.E.	1873
Reilly, Joy R.	B.S.	1890
Reinholdt, K. Oake P.	C.E.	1890
Rementer, George L.	C.E.	1884
Renshaw, Alfred H.	C.E.	1883
Reynders, J. V.W.,Jr.	C.E.	1886
Rice, Dan, Jr.	C.E.	1892
*Rice, Joseph G.	C.E.	1858
Rice, L. Frederic.	C.E.	1858
Rice, Spencer V.	C.E.	1871
Richardson, Harry L.	C.E.	1875
Ricketts, Palmer C.	C.E.	1875
Rickey, J. W.	C.E.	1894

Name.	Degree.	Class.
*Riddell, John L..A.B. (r.s.)		1829
Rider, J. B.....B.N.S., CE.		1844
Rider, Joseph B......C.E.		1889
"Rider, Thomas B....C.E.		1845
Ridgely, William B....C.E.		1879
Ripley,Thos. C...A.B.(r.s.)		1828
Roberts, George...... B.S.		1888
Roberts, G. B..C.E., B.N.S.		1849
Roberts, P....B.N.S., C.E.		1846
*Robison, John A...B.N.S.		1838
*Rocha, A. F. da......C.E.		1891
*Rockenstyre, Porter..C.E.		1849
Rockwood, Arthur J..C.E.		1887
Rockwood, C. F......C.E.		1894
Roebling, Charles G..C.E.		1871
Roebling, W. A.......C.E.		1857
Roebling, John A.....C.E.		1888
*Rogers, Horace N...C.E.		1837
Rood, Henry M.......C.E.		1885
*Root, Bennet.F..A.B. (r.s.)		1826
*Ropes, Charles F....M.E.		1871
Rosa, George de La...C.E.		1886
Rosa, Luis, de La....C.E.		1885
Rosenberg, Friedrich..C.E.		1882
*Rossman, Augustus..C.E.		1847
Rothwell, Richard P..C.E.		1858
Rousseau, Harry H...C.E.		1891
*Rowland, Frank L....C.E.		1875
Rowland, Henry A....C.E.		1870
Roy, Charles P.......C.E.		1893
Roy, Lawrence....... C.E.		1891
Royce, Harrison A....C.E.		1859
Ruggles, Charles H...C.E.		1892
Ruple, C.P...........C.E.		1881
Russell, Nathaniel E..C.E.		1870
Sabbaton, F.A.......C.E.		1892
Sabin, Alpheus T.....C.E.		1878
*Sage, Russell, 2d.... C.E.		1859
*Sager, Abram..A.B. (r.s.)		1831
Salisbury,James H..B.N.S.		1846
Salles, Joaquim de....C.E.		1879
Saltar, John, Jr......C.E.		1867
Samper, Julio........C.E.		1880
Sanders Francis N....C.E.		1891
*Sanders,W. S...A.B. (r.s.)		1833
Sanderson, Edwin N..C.E.		1886
Sanderson, J. G......C.E.		1858
*Sandford, Edw..A.B. (r.s.)		1827
Sariol, Pompeyo......C.E.		1867
Saulles, Arthur B. de..B.S.		1859
Sax, Percival M......C.E.		1890
Saylor, Francis H.....C.E.		1867
Scarborough, F. W....C.E.		1888
Schade, Charles G....C.E.		1892
Schaeffer, John S.....C.E.		1866

Name.	Degree.	Class.
Schermerhorn, R.....C.E.		1871
*Schott, C. Ridgely...C.E.		1868
Schultze, Paul L......C.E.		1891
Scott, Charles H......M.E.		1870
Searles, William H....C.E.		1860
Sedley, Henry........C.E.		1848
Selden, Samuel F.....C.E.		1886
Seligman, Albert J....C.E.		1878
Seminario, Juan......C.E.		1878
*Serrano, Aurelio.....C.E.		1860
Shankland, Edw. C....C.E.		1878
Shannahan, J. N......C.E.		1894
Sharp, William G.....C.E.		1879
Shaw, Henry C.......C.E.		1876
Shaw, Richard E......C.E.		1878
Sheal, Robert E.......C.E.		1894
Sheffield, John........C.E.		1891
Shepherd, Willard F..C.E.		1878
Sherman, William B..C.E.		1872
Shererd, Morris R....C.E.		1886
*Sherrill, Rush..A.B. (r.s.)		1830
Shields, Howard H....B.S.		1886
Shields, James W......C.E.		1890
Sikes, George R......C.E.		1886
Silliman, Justus M....M.E.		1870
*Simpson, B. V.......C.E.		1879
Simpson, William S...B.S.		1860
Singer, Robert R......C.E.		1877
Skilton, George S.....C.E.		1868
Skilton, James A.. .B.N.S.		1845
Skilton, Julius A....B.N.S.		1849
*Slade, Israel..C.E.,B.N.S.		1836
Slagle, W. C. H......C.E.		1892
Sloan, Robert I......C.E.		1859
*Small, T. B..B.N.S., C.E.		1843
*Smalley, D. S..B.N.S,C.E.		1835
Smith, Charles E......C.E.		1860
Smith, Charles R.....C.E.		1878
Smith, David C..A.B. (r.s.)		1833
Smith, Felix R. R.....C.E.		1860
Smith, Frank G.......C.E.		1859
Smith, H. DeWitt....C.E.		1875
Smith, Harmon M.... C.E.		1892
Smith, Milo A........C.E.		1867
Smith, Pemberton.....C.E.		1888
Smith, S. Kedzie......C.E.		1886
Smith, T. Guilford....C.E.		1861
Smith, Thaddeus S....C.E.		1861
*Smith, Theo. S., Jr...C.E.		1868
Snyder, Henry R......C.E.		1875
Sooysmith, Charles. ..C.E.		1876
Sosa, Pedro J........C.E.		1873
Sothers, Edward......M.E.		1870
Spearman, Francis....C.E.		1884
Springer, Lewis H....C.E.		1887

Name.	Degree.	Class.
Squires, John	C.E.	1869
*Stanton, L., Jr.	C.E.	1841
Starbuck, George H.	C.E.	1840
Starr, Arthur B.	C.E.	1869
Stearns, George A.	C.E.	1849
Stearns, Irving A.	M.E.	1868
Stebbins, O.	C.E., B.N.S.	1839
Steinacker, Theodore.	C.E.	1873
Stevenson, Holland N.	C.E.	1866
*Stevenson, P. E.	A.B.(r.s.)	1830
Stilson, William B.	C.E.	1867
Stites, Archer C.	C.E.	1887
Stribling, Ben A.	C.E.	1886
Stodder, George T.	C.E.	1863
Stone, Cyrus R.	C.E.	1867
Stone, Lowell H.	C.E.	1869
Storrs, Abel	A.B., (r.s.)	1831
Storrs, Arthur H.	C.E.	1883
Stowell, Charles F.	C.E.	1879
Stowell, Ellery	C.E.	1872
*Stratton, Norman.	C.E.	1838
Strawbridge, W. C.	M.E.	1870
Stuart, Alfred A.	C.E.	1879
Stutzer, Herman, Jr.	C.E.	1878
*Suffern, Edward	C.E.	1835
Sugden, Clarence H.	C.E.	1889
Sutermeister, A. H.	C.E.	1892
Sutherland, Mosher A.	C.E.	1861
*Sutherland, Sam. W.	C.E.	1846
Swift, Alexander J.	C.E.	1872
Sykes, George W.	C.E.	1893
Symington, Wm. N.	C.E.	1861
Taylor, Gil. T.	C.E., B.N.S.	1844
Thacher, Edwin	C.E.	1863
Thackray, George E.	C.E.	1878
*Thomas, Jos.	A.B.(r.s.)	1830
Thomas, Samuel R.	C.E.	1891
Thomas, William H.	C.E.	1891
*Thompson, A. A.	B.N.S.	1838
Thompson, Arthur W.	C.E.	1892
*Thompson, Chas. B.	B.S.	1860
Thompson, Clark W.	C.E.	1887
Thompson, E. Ray	C.E.	1876
Thompson, Jas. G.	B.N.S.	1848
*Thompson, John C.	C.E.	1865
Thompson Mackey J.	C.E.	1893
Thompson, William A.	C.E.	1869
*Thomson, James P.	C.E.	1888
*Tibbits, George.	C.E.	1849
Tiernan, Austin K.	C.E.	1894
Tilghman, J.	B.N.S., C.E.	1839
Tompkins, Daniel A.	C.E.	1873
Tompkins, John A. B.	C.E.	1879
Tone, Sumner L. R.	C.E.	1886
Torkington, Isaac.	C.E.	1887

Name.	Degree.	Class.
Touceda, Enrique	C.E.	1887
Townsend, John.	C.E.	1879
*Trafton, Gilman.	C.E.	1856
Travell, W. B.	C.E.	1894
Trevor, Frank N.	C.E.	1866
*Trujillo, Francisco.	C.E.	1857
Tullock, Seymour W.	C.E.	1877
Tumbridge, John W.	C.E.	1891
*Tuomey, Michael.	B.N.S.	1835
*Turknett, Robert G.	C.E.	1886
*Turner, Benjamin.	C.E.	1849
Turner, Bejamin M.	C.E.	1888
Turner, Daniel L.	C.E.	1891
Tuttle, Frank W.	C.E.	1878
Tweeddale, William.	C.E.	1853
Ubsdell, John A., Jr.	C.E.	1889
Underwood, John C.	C.E.	1862
*Underwood, J. R.	C.E.	1875
Uribe, German.	C.E.	1893
Utley, Charles H.	M.E.	1869
Van Bergen, R. H.	C.E.	1841
Van Buren, John D.	C.E.	1860
Van Buren, Robert.	C.E.	1864
Van Hoesen, E. F.	C.E.	1878
*Van Ness, S.	C.E., B.N.S.	1836
*Van Rensselaer, A., A.B.(r.s.)		1833
Van Rensselaer, P., B.N.S.,C.E.		1839
*Van Schaick, A. P.	C.E.	1839
Van Sinderen, A., B.N.S.,C.E.		1847
Van Zile, Harry L.	C.E.	1884
Varona, Ignacio M. de.	C.E.	1863
Vaughan, Edgar.	C.E.	1894
*Vaughan, F. W.	C.E.	1863
Verner, Henry W.	C.E.	1881
Verner, Morris S.	C.E.	1876
Vier, Henry.	C.E.	1883
Viscarrondo, L. J. de.	C.E.	1859
*Voorhees, Herman.	C.E.	1873
Voorhees, Paul.	C.E.	1884
Voorhees, Theodore.	C.E.	1869
*Vought, William G., C.E., B.N.S.		1840
Vroom, Peter D.	C.E.	1862
Waddell, John A. L.	C.E.	1875
Waddell, Montgomery.	C.E.	1884
Wade, James, Jr.	C.E.	1842
Wagner, Richard G.	C.E.	1887
Wainwright, J. T.	C.E.	1875
Waite, Christopher C.	C.E.	1864
Waite, Guy B.	C.E.	1888
Walbridge, Russell D.	C.E.	1871
Walbridge, T. Chester.	C.E.	1873

Name.	Degree.	Class.	Name.	Degree.	Class.
Walbridge, Thos. H.	C.E.	1876	Williams, Clifton G.	C.E.	1877
Walbridge, W. G.	C.E.	1877	*Williams, James B.	C.E.	1888
Walker, William W.	C.E.	1856	*Williams, J. Francis	C.E.	1883
*Walker, William W.	C.E.	1886	*Williams, Norman A.	C.E.	1859
Wallace, Gurdon B.	C.E.	1840	Williams, Samuel W.	C.E.	1894
Wallace, James P.	C.E.	1837	*Williams, S. Wells,		
Wallace, William M.	C.E.	1892		A.B.(r.s.)	1832
Waller, William, Jr.	C.E.	1879	Williams, Theodore H.	C.E.	1889
Walsh, George S.	C.E.	1894	*Williams, W. B.	C.E.	1835
Walter, Alfred	C.E.	1872	Williamson, T. M.	M.E.	1871
Waltz, Joseph E.	C.E.	1877	Willson, Fred N.	C.E.	1879
Ward, Vincent B.	C.E.	1886	Wilson, Henry W.	C.E.	1864
Ware, R. Willard	C.E.	1850	Wilson, Howard M.	C.E.	1884
*Warren, Levi H.	C.E.	1837	Wilson, James M.	C.E.	1887
Warren, Ogle T.	C.E.	1891	Wilson, John A.	C.E.	1856
Warren, S. Edward	C.E.	1851	Wilson, Joseph M.	C.E.	1858
*Watkins, Hezekiah.	C.E.	1857	Winger, Oswald E.	C.E.	1886
Watriss, George C.	C.E.	1853	Winslow, Charles W.	C.E.	1858
Weir, Chas. G.	C.E.	1877	Witmer, Joseph F.	C.E.	1887
Wellington, Geo. B.	C.E.	1875	Witmer, Victor M.	C.E.	1887
Wells, Joseph A.	C.E.	1883	Wood, Charles W.	C.E.	1884
*Westcott, A.	C.E., B.N.S.	1835	Wood, De Volson	C.E.	1857
*Weston, C. L.	A.B.(r.s.)	1827	Woodruff, Joel R.	C.E.	1848
Weston, Charles S.	C.E.	1882	*Woodward, F. G.	C.E.	1839
Wheeler, Fred L.	C.E.	1894	Woodworth, B.B.	A.B.(r.s.)	1833
*Whipple, Charles	C.E.	1837	*Woodworth, John, Jr.	C.E.	1837
*Whipple, Stephen T.	C.E.	1838	Worcester, Geo. W.	C.E.	1887
Whistler, Thomas D.	C.E.	1881	Worthington, Charles.	C.E.	1892
White, Alfred T.	C.E.	1865	*Wotkyns, A. A.,		
*White, John H.	C.E.	1840		C.E., B.N.S.	1847
Whitner, James H.	C.E.	1885	Yardley, Edmund	C.E.	1856
Whitney, Drake	C.E.	1864	Yates, Preston K.	C.E.	1880
*Whittelsey, P.D.	A.B.(r.s.)	1834	Yeager, Frederick A.	C.E.	1878
Wigand, Albert A.	C.E.	1889	Young, Don Carlos	C.E.	1879
Wiggins, Charles, Jr.	C.E.	1878	*Young, Feramorz L.	C.E.	1879
Wilde, N. R.	C.E., B.N.S.	1836	Young, Frederick S.	C.E.	1880
Wiley, William H.	C.E.	1866	Young, Horace G.	C.E.	1877
Wilkins, William G.	C.E.	1879	Young, Jonas F.	C.E.	1872
*Wilkinson, Alfred	C.E.	1849	Zabriskie, Aaron J.	C.E.	1876
*Wilkinson, J. F.	C.E.	1847	Zayas, Octavius A.	C.E.	1886
*Wilkinson, W.	A.B. (r.s.)	1830	Zegarra, Enrique C.	C.E.	1874

INDEX.